Switched Parasitic Antennas
for Cellular Communications

Disclaimer of Warranty

The technical decriptions, procedures, and algorithms in this book have been developed with the greatest care; however, they are provided without a warranty of any kind. Artech House, Inc., and the authors of the book entitled *Switched Parasitic Antennas for Cellular Communications* make no warranties, expressed or implied, that the equations and procedures in this book are free from error, or will meet your requirements for any particular application. They should not be relied on for solving a problem whose incorrect solution could result in injury to a person or loss of property. The authors and publisher disclaim all liability for direct, incidental, or consequential damages resulting from your use of the procedures in this book.

For a listing of recent titles in the *Artech House Antennas and Propagation Library,* turn to the back of this book.

Switched Parasitic Antennas for Cellular Communications

David V. Thiel
Stephanie Smith

Artech House
Boston • London
www.artechhouse.com

Library of Congress Cataloging-in-Publication Data
Thiel, David V.
    Switched parasitic antennas for cellular communications / David V. Thiel,
    Stephanie Smith.
       p.   cm.—(Artech House antennas and propagation library)
    Includes bibliographical references and index.
    ISBN 1-58053-154-7 (alk. paper)
    1. Adaptive antennas.   2. Switching circuits.   3. Radio—Interference.
    4. Cellular telephone systems.   I. Smith, Stephanie, 1974–   II. Title.   III. Series.
    TK7871.67.A33   T47   2001
    621.384'135—dc21
                                                  2001045736

British Library Cataloguing in Publication Data
Thiel, David V.
    Switched parasitic antennas for cellular communications.—(Artech House antennas
    and propagation library)
    1. Antennas (Electronics)   2. Wireless communication systems
    I. Title   II. Smith, Stephanie
    621.3'824
    ISBN 1-58053-154-7

Cover design by Gary Ragaglia

International Standard Book Number: 1-58053-154-7
Library of Congress Catalog Card Number: 2001045736

10 9 8 7 6 5 4 3 2 1

Contents

Preface

Since the early 1990s, a large number of books about antennas have become available, both new and updated editions. It is not the aim of our book to compete with these comprehensive texts, which cover almost every aspect of antenna design. To our knowledge, however, none discusses switched parasitic antennas in as much detail as we do here. This book introduces the basic concepts and applications of this antenna class, which has many potential applications across a broad range of communications areas, particularly with the introduction of smart antennas to the cellular communications field. These antennas have the advantages of relatively inexpensive design and building costs, sturdy construction, computer control, high-speed direction finding, and a small antenna volume.

In introducing this relatively new technology, we first provide background on basic antenna theory. Both wire and patch antenna structures are discussed, together with their circuit requirements. We hope to offer the reader sufficient guidance to undertake the design and implementation of a switched parasitic antenna.

We have attempted to discuss the basic antenna concepts in a clear and straightforward way, by reducing the antenna descriptors to a small number of parameters: input impedance defined by the scattering parameter S_{11}, the beamwidth (as a representation of the total polar pattern), the bandwidth, and gain. Many experienced antenna engineers might find this approach limiting and less than rigorous, but we believe that all of the other descriptors such as antenna directivity and effective radiated power are extensions of these basic parameters. Certainly the parameters we use are sufficient to include in the cost function of an optimization procedure to finalize the antenna design.

The chapter on optimization techniques is unusual in an antenna book, even though antenna structures are rarely simple and so require multi-parameter optimization techniques to achieve a desired goal. With the advent of many computer techniques to solve the forward problem (i.e., calculation of the response of the antenna given the antenna description), the inverse problem (i.e., calculation of the antenna description given the antenna properties) can be solved using a combination of an optimization routine and an efficient solver. The data resulting from this process also yields a sensitivity analysis that directly impacts on the manufacturability of the final design—that is, the tolerances of each of the design parameters. In the next few years, computational engines will continue on the path of increasing speed and greater memory, enabling more complex optimization problems to be undertaken. But lest educators discard the subject of antenna design and companies fire their antenna design engineers, it must be stressed that knowledge of the basics of antenna theory and design is essential for antenna optimization. Skill and background experience remain key elements of the overall design process.

We would judge the value of this book on the ability of a reader to successfully design smart antennas based on a switched parasitic antenna using either wire or patch elements. We would appreciate any comments from readers about the text's effectiveness based on this criterion.

Acknowledgments

Many people assisted us in the development of this book. Some years ago, a reporter asked David Thiel how long it took him to develop a particular antenna. David knew that the answer to this question had to take into account the time elapsed since his first course in electromagnetics! Even before that, the encouragement of parents and family, friends, and teachers contributed to the development of the two individuals who wrote this book.

We authors would like to thank the past and present members of the Radio Science Laboratory at Griffith University. In particular, discussions over the years with Steven O'Keefe, Jun Wei Lu, and postgraduate students Derek Gray, Seppo Saario, Gregory Durnan, Robert Schlub, and research assistant Andy Byers, and many undergraduate students who became enthused about antennas, who asked questions of every variety, and challenged us to rethink our explanations about why things happened.

David would particularly like to thank Inari for her patience over the years as he developed his photographic portfolio of "Antennas I Have Known."

Stephanie would like to thank CSIRO Telecommunications and Industrial Physics for their assistance in building and testing some of the antennas presented in this book, with special thanks to Trevor Bird, one of her mentors. Thanks also to Stephanie's husband Trevor, for his comments and editing of this book.

David V. Thiel
Stephanie Smith
Australia
October 2001

1

Smart Antennas

1.1 Introduction

Cellular telephone systems are popular worldwide, and the number of users continues to rise rapidly. This has resulted in pressure on the existing cellular network and a drive to develop a more efficient network based on the current spectrum allocations. Research has focused on the use of smart-antenna technologies in which identical frequencies can be used in the same cell, increasing the maximum number of users in the cell without the allocation of additional radio frequency bands. This book examines one approach to smart antennas, with particular applications to cellular telephone technology.

In the mid-1990s the terms "smart antenna" and "intelligent antenna" were introduced to describe antenna systems with control of the radiation characteristics—mainly direction—to tackle a number of important problems in wireless communications [1–4]. These problems include:

- Multipath fading, where the signal arrives at the receiver from a number of different directions and with a number of different time delays;
- Electromagnetic compatibility (EMC), where other transmitters in the vicinity degrade the communications network, and the presence of this communications network adversely affects the performance of other nearby systems;

- Frequency reuse, where the same frequency is used in more than one communications link in the same location.

All of these problems result in some form of degradation of the communications link. One obvious solution is to use directional antennas. This solution is not particularly flexible if the communications network is designed for mobile users or when the communications network is upgraded on a regular basis. For example, upgrades are required when the user capacity of a cellular telephone network is exceeded and additional intermediate cell sites must be installed, or when the physical propagation path is altered due tobuilding construction or the presence of a new interference source. An adaptive antenna system that is controlled electronically has advantages. A number of books have been dedicated to recent advances in antenna technology designed specifically for communications systems [5, 6].

Four broad classes of antennas can be controlled electronically. These are:

1. Electromechanical antennas, where the physical orientation of the antenna structure is controlled by electric motors;

2. Phased arrays, where the radiation from an array of identical elements is controlled by electronically changing the phase path to each element;

3. Active integrated antennas, where the phase of the signal to each element of a large array is controlled electronically by two integrated amplifiers;

4. Switched parasitic antennas, where the direction of the radiation from the antenna is controlled by electronically changing the characteristics of nearby parasitic elements close to the driven element.

The basic function of a transmitting antenna is to convert electrical signals from a radio frequency (RF) generator into an electromagnetic wave directed toward either a receiver or a target. A receiving antenna converts an incoming electromagnetic wave to an electrical signal, preferably with the exclusion of all other electromagnetic signals in the vicinity. In both cases, this functionality is achieved by designing the antenna to have directionality and frequency selectivity. In this book, the directionality of an antenna is described in terms of the principal beam direction and beamwidth, and the frequency selectivity is described in terms of the bandwidth of the antenna.

Antennas having their beam direction and bandwidth controlled electrically or electronically are of significant interest to the telecommunications industry. Before the cellular telephone revolution began in the late 1980s, controlled antennas were principally used for direction-finding applications such as radar and satellite tracking. Most of these antennas were electro- mechanically operated. The expanding demand for wireless communications started with voice-only communication systems in the form of portable wireless telephone handsets for the mass markets around the world. There is increasing pressure on the finite electromagnetic spectrum with the inclusion of other forms of communications. These include the transmission of sound and images through electronic mail (e-mail), electronic commerce (e-commerce), and the Internet. This has resulted in communications engineers seeking frequency reuse in closely spaced cells, in addition to solving the better known problems of wireless communications such as multipath fading, electromagnetic interference (EMI), and electromagnetic compatibility (EMC). These problems can be alleviated by the use of smart-antenna systems.

Delisle et al. [7] suggested that for an antenna to be "smart" or "intelligent," a computing system is required. They divided smart antennas into two classes—switched-beam antennas and adaptive-beam antennas. Switched-beam antennas are antennas in which only the direction of the maximum radiation is controlled. Adaptive-beam antennas are antennas in which the radiation in all directions is controlled simultaneously. For example, it is possible to direct a null in the radiation pattern toward an interfering source, and maintain the direction of the maximum radiation at the same primary source when both source and interferer are moving. In both cases, only antennas with more than one element driven with the RF signal were considered. In this book, a system of smart or intelligent antennas is described in which only one element of the antenna is active, and so the problems associated with multiple active elements, such as high cost, significant radio frequency power loss caused by the use of power splitters and phase shifters, and radiation from a complex transmission line feed structure, are either minimized or eliminated completely. With only one driven element, only one principal beam direction is possible at any time.

If the antenna system can be used in more than one direction simultaneously (i.e., the antenna has more than one principal beam at any time), then the number of active elements must be at least equal to the number of simultaneous beams. This allows the antenna system to receive transmissions from more than one direction at the same frequency simultaneously. This type of antenna system provides spatial-division multiple access (SDMA). In most cases, the antennas discussed in this book will provide only angular diversity through a single steerable beam. SDMA can be provided by an antenna with more than one beam or by using more than one antenna in the same cell.

In many cases it is desirable to have an antenna system with a wide bandwidth that can be used in more than one frequency band simultaneously. For example, in mobile satellite communications, the uplink center frequency is usually significantly different from the downlink frequency, yet it is preferable to have only one antenna for both the uplink and the downlink channels. In cellular telephone communications, for example, there is a move to more flexible systems, with the development of a single handset that can operate in a number of different frequency bands, for example, GSM at 960 MHz and PCS at 1,900 MHz. In a small portable device, an array of antennas is expensive in cost, weight, and space, so the use of a single antenna structure is an advantage. At the cellular telephone base station also, it is sometimes necessary to operate a number of different communications systems using different frequencies with the same antenna.

From an economic point of view, there is always a need to reduce the size of the antenna arrays, as this usually results in a reduced fabrication cost and a lower cost for the installation support structure. In addition, the generally negative impact on the visual environment (the aesthetics of the antenna) can be reduced.

The design of these antenna systems must include the following technical and aesthetic considerations:

- Beamwidth;
- Bandwidth;
- Input impedance;
- Multifrequency capability;
- Size and weight (including footprint area);
- Speed of operation.

These issues are addressed in this book. Switched parasitic antennas offer some improvements over current switched-beam- and adaptive-beam-antenna technologies. Initially, we will discuss basic antenna and control definitions and concepts that serve to emphasize why these parameters are important. This establishes a framework for comparing the performance of different antenna systems.

1.2 Tracking Radio Sources

For both clarity and simplicity, we will describe the smart antenna in its function as a stationary receiver. A mobile transmitting system is used to complete the communications link. This allows the origin of the coordinate system to be set at the center of the smart antenna. Other arrangements require a shift in the origin from this coordinate system. This is not difficult to do in principle. As a consequence of the principle of reciprocity for antennas, the radiation pattern of an antenna is identical for both transmission and reception if surface waves are excluded. In some cases it is easier to discuss the radiation characteristics in terms of transmission, and in other cases in terms of reception. These concepts are well explained in most basic electromagnetic and antenna textbooks [8–11].

Figure 1.1 illustrates a number of basic wireless communications systems. The most common view of a communications link includes the initiator of the communication and the responding system. Both terminals are capable of two-way (i.e., bidirectional or duplex) exchange of information using electromagnetic (EM) waves [Figure 1.1(a)]. In this book, the terms "communications link" or "communications system" will be taken more generally to include radar systems and direction-finding systems, as well as this narrower definition of communications. The differences between radar, passive direction finding, and more standard communications systems are not great. For example, in a simple monostatic radar system, the transmitter and the receiver use the same antenna to send and receive EM signals with the information contained in the reflection from the target object [Figure 1.1(b)]. In a bistatic radar system, the receiving antenna is different from the transmitting antenna and again the information content of the transmission lies in the reflection from the target [Figure 1.1(c)]. In a simple direction-finding system, the base station is a receiver only, and the information content is the characteristic EM emanations from the transmitter [Figure 1.1(d)]. All of these systems require at least some of the basic building blocks shown schematically in Figure 1.1(e). The frequency of the RF source is referred to as the carrier frequency, and the frequency of the information is referred to as the base-band frequency.

In more sophisticated radar systems, a cooperative target has a transponder. On detection of the incoming radar pulse, the target responds on a different frequency with an identification signal. This allows the radar system to track multiple targets simultaneously without the need for sophisticated radar cross section information to achieve target recognition.

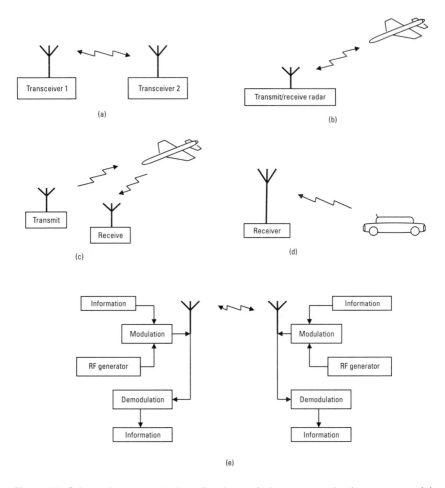

Figure 1.1 Schematic representation of various wireless communications systems: (a) basic communications link; (b) monostatic radar system; (c) bistatic radar system; (d) direction-finding system; and (e) generalized communications system.

A smart antenna must be capable of tracking an RF source as it moves through its field of detection. This depends on the directionality of the antenna beam, the detector sensitivity, and the RF field strength of the source. There are three possible mobile communications configurations:

1. The source is mobile and the receiving antenna is stationary (e.g., a ground station tracking a polar-orbiting satellite; a cellular

telephone base station communicating with a moving person or vehicle; a ground-based radar system at an airport tracking incoming aircraft).

2. The receiving antenna is mobile and the source is stationary (e.g., a moving vehicle communicating with a geostationary satellite; the handset of a cellular telephone; a radar on a boat scanning for a navigation beacon).

3. Both the transmitter and the receiver are moving (e.g., a portable handset communicating with a polar-orbiting satellite; a radar system mounted in a moving vehicle used to detect the location of other moving vehicles).

From a user's perspective, an important property of an electronically controllable antenna is the tracking time required, both in the initial acquisition of a signal and in the maintenance of the communications channel once acquisition has been achieved. To track a moving source, it is necessary to check the signal level at a number of different antenna directions and select the direction where the incoming signal either has maximum field strength or the signal-to-noise ratio (SNR) is maximized. During this search procedure, the received signal into the demodulator is less than optimal, and information during this period may be lost. A fast and efficient tracking system will ensure the rapid acquisition of the signal from the moving source. If the time required for the search process is minimized, the signal into the demodulator will be optimal over a longer period of time.

Two timing parameters are used to describe the tracking system: the acquisition time and the dither time. The acquisition time t_a is the time required for a global search of all possible directions to locate a signal. The dither time t_d is the time required for a local search where only those directions adjacent to the most recent position of maximum signal strength are scanned. The position of maximum signal is updated after each local search. These two parameters t_a and t_d can be used as a basis for comparison of all electronically controllable communications systems. They depend on a number of factors, which include:

- The beamwidth of the antenna (B_θ and B_ϕ);

- The number of discrete orientation positions of the main beam (N_{tot});

- The number of beams that the antenna can employ simultaneously (N_a);
- The decoding time required for signal identification (t_c);
- The computation time (t_o) required by the computing system to verify the identity of the signal and to update the position-control functions of the antenna to direct the beam accordingly.

We will now discuss these factors in more detail.

1.2.1   Beamwidth

The beamwidth of an antenna is defined in terms of the range of angles over which the antenna has maximum sensitivity. If the beam can be controlled in one angular direction only, say, the ϕ plane, then the control of the beam is labeled one-dimensional (1D). If the beam can be controlled in two angular directions, the θ and ϕ directions, then the control of the beam is two-dimensional (2D). The angles θ and ϕ are defined in a spherical coordinate system shown in Figure 1.2, where the xy plane is parallel to the local surface of the Earth (the horizontal plane). The 1D beamwidth B_ϕ, is defined as the angular range of the main beam of the antenna that lies within −3 dB of the power in the direction of maximum gain of the antenna. This is the half-power width of the radiation pattern in the ϕ direction. In the 2D case, the −3-dB direction is defined as a cone with elliptical cross section, with two orthogonal half-power arcs B_θ and B_ϕ, defining the extent of the ellipse as indicated in Figure 1.2.

The selection of the −3-dB level to define the beamwidth is somewhat arbitrary. In most cases, the dynamic range of the receiving system will be capable of much lower-level input signals. In order to minimize EMI in the vicinity of a transmitting antenna, and to reduce interference from adjacent channels, many communications systems maintain tight control over the radiated power level of both ends of a communications link. For this reason, the selection of −3 dB rather than a lower value (often −10 dB) is usually used as the best compromise.

1.2.2   Main Beam Scanning

The direction of the main beam or the direction of the principal radiation is defined in terms of θ and ϕ. An antenna with no electronic control will have a constant principal radiation direction.

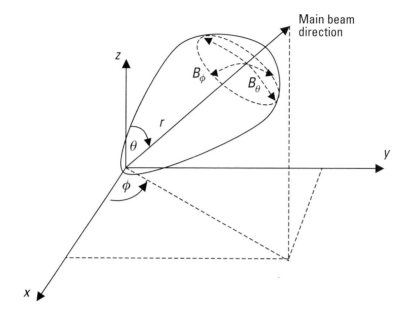

Figure 1.2 Coordinate system used to define the direction of the main beam of an antenna and the beamwidths in the two planes B_ϕ and B_θ.

In an electronically controllable antenna, the direction of the main beam is adjusted by the application of electrical signals to the active devices in the antenna system. These devices are usually RF switches, RF phase shifters, or RF amplifiers. In switched parasitic antennas, switches are used to control the resonant frequency of the elements, both active and parasitic, or alternatively to redefine the active element within an array of elements. In phased arrays, switches are used to control the phase shift in a transmission line feeding a particular element [12]. Assuming that the controller of the antenna is a digital computer system, the control of the antenna is discrete rather than continuously variable. Another alternative is to use RF amplifiers to control the amplitude and phase of the RF signal delivered to the antenna [13].

The number of discrete switching positions of the elements in the antenna system, both active and passive, determines the number of different positions of the main beam. The controller requires the same finite number of switching states. There is little advantage in controlling the direction of the main beam of the antenna at angles more closely spaced than one beamwidth in either angular direction. These discrete angles can be defined in terms of the θ and ϕ positions shown in Figure 1.2. The number of beam positions in each angular range, N_θ and N_ϕ, are the number of discrete beam positions in the θ and ϕ directions, respectively.

For a ground-based antenna system, the maximum range of angles for θ and ϕ are given by the inequalities $0 \le \theta < \pi / 2$ and $0 \le \phi < 2\pi$. The relationship between the number of beam positions and the beamwidth in the two directions must be constrained by the inequalities:

$$N_\theta \le 1 + \pi / 2B_\theta \text{ and } N_\phi \le 2\pi / B_\phi \qquad (1.1)$$

The technique of using equally spaced steps in a uniform sequence is referred to as sequential uniform sampling. It is tempting to assume that the total number of positions N_{tot} required for a complete scan of all possible angles is given by:

$$N_{tot} = N_\theta N_\phi \qquad (1.2)$$

For coverage of the complete hemisphere, it is evident from Figure 1.2 that when the beam is pointing directly vertical along the z-axis (i.e., for $\theta = 0$), and the beam is symmetrical about this axis, the beam position is independent of ϕ.

Equation (1.2) is incorrect because the number of independent ϕ steps necessary for a full rotation decreases as θ decreases. The concept of the solid angle [9] is introduced to calculate a more accurate value for N_{tot}. The solid angle of an antenna Ω_a is defined in terms of the footprint of the -3-dB ellipsoid on the area of an imaginary sphere centered on the antenna. The shape is not a true ellipse, as the projection is onto the surface of a sphere. Figure 1.3 shows that if the radius of the imaginary sphere is r, then the beamwidth B_θ creates an arc of length rB_θ on the surface of the sphere, and the beamwidth B_ϕ creates an arc of length rB_ϕ. These two lengths define the area of the -3-dB ellipsoid that constitutes the angular spread of the principal direction of the main beam of the antenna.

The solid angle Ω_a is defined as the area of the -3-dB ellipsoid A_B on the sphere relative to the radius squared, that is,

$$\Omega_a = A_B / r^2 \qquad (1.3)$$

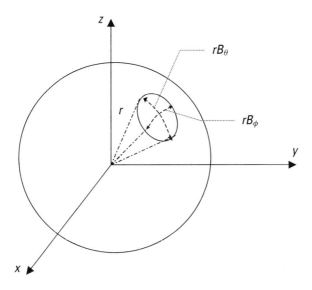

Figure 1.3 Beamwidth ellipse projected onto the surface of a sphere of radius *r*, centered on the antenna.

For an omnidirectional radiation source (i.e., one that radiates in all directions with the same intensity), the sphere is covered totally by the radiated field and the maximum solid angle $\Omega_a = 4\pi$ is obtained. Given that the –3-dB beamwidths are B_θ and B_ϕ, and the projection onto the surface of a sphere is approximately an ellipse, the area of this ellipse A_B is approximately given by:

$$A_B \cong \pi B_\phi r B_\theta r \,/\, 4 \tag{1.4}$$

If the maximum scan range is a complete hemisphere, then the total area to be covered in one global scan is $2\pi r^2$ and the total number of scan positions N_{tot} required to cover the complete area is approximately:

$$N_{tot} \cong 8\pi r^2 \,/\, (\pi B_\theta r B_\phi r) \cong 8 \,/\, B_\theta B_\phi \tag{1.5}$$

This equation can be written in terms of N_ϕ and N_θ by substitution of the expressions from (1.1):

$$N_{tot} \cong \frac{8N_{\phi}(N_{\theta} - 1)}{\pi^2} \tag{1.6}$$

This approximation has the underlying assumption that B_{θ} and B_{ϕ} do not change over the hemisphere, and the individual beams can be packed together with maximum density. The second is not true, as the individual beam ellipsoids must overlap for total area coverage. Equation (1.6) gives the minimum value for the number of beam positions required to complete a hemispherical scan if the −3-dB beamwidth is rigorously applied.

1.2.3 Multiple Beam Scanning

A more flexible arrangement for an antenna system is one that can track more than one transmitter at a time. This concept is relatively new, with most smart-antenna systems having a single principal direction. Before the development of this multibeam concept, some adaptive antennas with a single principal direction employed additional control software to position the null of the main beam at the angular location of an interference source or a second path from the desired transmitter. In this case, even if there is some sacrifice in the gain of the main lobe, the reduction of an interfering source may improve the communications link. The possibility of a single antenna system with multiple communications channels active simultaneously has been discussed recently. As this can be achieved using phased arrays, integrated active antennas, and switched parasitic antennas, this possibility is investigated briefly.

If the number of simultaneously active beams for a single antenna is N_a, then the number of beam positions required in a global scan is reduced. There is no advantage in providing two beams in precisely the same direction, so the total number is reduced by N_a. For two-dimensional hemispherical scanning using (1.6), N_{tot} becomes:

$$N_{tot} \cong \frac{8N_{\phi}(N_{\theta} - 1)}{\pi^2} - N_a + 1 \tag{1.7}$$

Here N_{tot} is the number of beam positions scanned when the other beams are already active. If only one beam is used, (1.7) reduces to (1.6).

In the case of an initial global search, one possible solution is to divide the hemisphere into sectors with equal solid angle, and use each beam to scan a particular sector. In this case, the number of beam positions required for each scan is:

$$N_{tot} \cong \frac{8N_\phi(N_\theta - 1)}{\pi^2 N_a} \tag{1.8}$$

Depending on the algorithm employed, this search pattern may restrict the antenna to one beam position for each sector. While this will lead to a missed signal source if two are present in the same sector, the scan time is significantly reduced using this method of approach.

1.2.4   Signal Identification Time

Communications systems carry information in the form of modulation. Such modulation may consist of a combination of amplitude modulation, frequency modulation, and phase modulation. In establishing a communications link, it is necessary to identify that the signal received is not from a spurious source. For example, in a simple radar system, the identification of a radar return can be made using a threshold determination. In this case the speed of determination is related to the settling time of the detector and the speed of the analog-to-digital converter, which presents the information to the controller. Typically, an RF detector will have a response speed of less than two cycles of the carrier frequency. The noise immunity in this case is quite poor, because any local noise of sufficient intensity at the radar operating frequency can trigger the receiver.

If the radar identification requires an echo that is phase related to the transmitter, then the use of a phase-locked loop is required. In this case, the response speed is limited by the lock-in speed of the phase-locked loop [14]. This is commonly of the order of several cycles of the carrier frequency, and depends on the type of phase-locked loop employed and the application of the radar.

In a channel using some form of coding—for example, spread spectrum [e.g., frequency-division multiple access (FDMA)]—or a wireless network with many nodes [e.g., code-division multiple access (CDMA), or time-division multiple access (TDMA)]—the system must be capable of detecting and correctly decoding the identifying pulse sequence to validate that the desired signal is incident from this direction. This introduces a delay in the signal identification time or coding time t_c. In these circumstances, a multiple sequence of bits at the base-band frequency must be received and identified. This, in turn, requires a large number of carrier frequency cycles for a positive identification to be made. The time t_c is dependent on the coding system employed, and will require a significant number of bits in the base-band frequency of the communications system.

1.2.5   Computation Time of the Controller

In an electromechanical system, the controller must be programmed to take into account the speed of the antenna drive motor, the torque generated by the motors to change the angular direction of the antenna, and the inertia of the antenna. In many applications, this time can be large in comparison with the delay times discussed earlier.

In a completely electronic antenna system, there is no physical movement of the antenna structure. The digital support electronics (e.g., computer, microcontroller, microprocessor, programmable logic device, gate array) is synchronized by a clock. The time taken between receiving an input, and the output of a control signal to switch the antenna, is related to the complexity of the computations involved. These computations generally perform three different functions. First, the signal is positively identified. Second, the signal level is compared with the signal levels received by adjacent beam positions. And third, the decision for the new/updated position is computed and enacted. Computing systems are now running at clock speeds above 500 MHz, and this continues to increase rapidly, so that the speed of these operations continues to increase. Whether for global searching or local dithering, the time taken to update the position precisely depends on the complexity of the algorithm. The total time required between signal identification and the output of the control function is referred to as t_o. The time required to scan all beam locations depends on $N_{tot}t_o$, the product of the total number of beam positions that will be used in global search mode and the computation time t_o. In dither mode, the number of beam positions required to be searched is much smaller and so the update time required is correspondingly smaller.

1.3   Antenna Control Systems

The function of the control system for a smart antenna is to direct the main beam for optimum communications performance. This usually means the largest SNR. The controller tasks are as follows:

1. Scan the field of view for a possible transmitter (a global search).
2. Identify that the transmitter is a valid target.

3. Set the beam direction for maximum SNR.

4. Store the location of maximum response in memory.

5. Recall the last position of maximum response from memory and check the SNR at adjacent directions (local search or dithering).

6. Update in memory the new position of the transmitter and use this together with (5) to predict the next update time and location.

Achieving these functions in the shortest possible time will ensure maximum information transfer through the communications and processing system. Figure 1.4 illustrates the basic building blocks required for the control system of a smart antenna when in receiving mode. The receiver demodulates the RF signal from the antenna, converting it to the base-band frequency, and provides two outputs: one for the communications terminal and the other for the antenna control system. The communications terminal makes an identification of the incoming signal and passes this information to the controller. The signal-level information is recorded and then compared with the signal level detected in other beam positions. The direction of the antenna beam is changed if the current direction is not optimal. This change is made by electrical signals to the switches or phase shifter electronics.

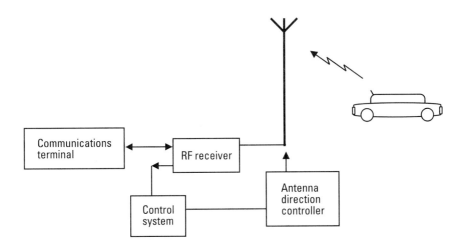

Figure 1.4 Functional blocks used in a smart-antenna system.

1.3.1 One-Dimensional Tracking

In order to explore how a tracking system operates, consider the case of an airport radar that has beam control in the ϕ direction only. Assume that the antenna has a beamwidth in the ϕ plane of $B_\phi = 0.035$ radians (2°), and that there is mechanical control of the beam. A further assumption is that the position of the beam in the ϕ direction is not continuous but rather consists of 180 equally spaced, discrete steps of 0.035 radians each (i.e., $N_\phi = 180$). Two cases are considered; in the first case, the rotation is possible in one direction only, and in the second, both directions of rotation are possible.

1.3.1.1 Case One

Consider the case of sequential uniform sampling in which the antenna rotates in the $+\phi$ direction about the z-axis at a constant angular speed of $\omega_\phi = 2\pi$ radians per second. For every updated position, all possible locations are checked for maximum SNR. If the position of the signal is first detected at angle ϕ_1, then the time taken to obtain the next position determination depends on the direction of rotation of the source. For example, if the next position of the source is $\phi_1 + B_\phi$, then the time taken for the antenna to be directed in this direction is B_ϕ/ω_ϕ. If the next position is at $\phi_1 - B_\phi$, then the time taken for the antenna to face this direction is $(N_\phi - 1)B_\phi/\omega_\phi$. Thus, an accurate determination is guaranteed only once per revolution of the antenna. For the rotation speed specified, the update time is once every second. This continuous global scan approach is the only method possible if the antenna is constrained to move in one angular direction only.

The angular position of the source is defined after the global search has been completed. The total time required, t_a, is given by the sum of the decoding time t_c, the orientation time t_o, and the rotation time required for each incremental step of one beamwidth B_ϕ, for each of the total number of steps N_{tot}. This can be written as:

$$t_a = N_{tot}(t_c + t_o + B_\phi / \omega_\phi) \tag{1.9}$$

1.3.1.2 Case Two

The time required for a successful identification of the signal can be reduced if the antenna can be rotated in both the $+\phi$ and $-\phi$ directions. Once the incoming signal has been detected, rather than completing a full 2π radian sweep, the antenna can be scanned incrementally to the left and right of the last position to gain an update on the direction of the incoming signal. This localized scan approach to direction finding is referred to as the dither mode.

If the antenna can rotate in both the $+\phi$ and $-\phi$ directions, the global search time t_a given in (1.9) is substantially reduced, providing that the mechanical inertia of the antenna structure can be overcome by the drive motors. This update time or dither time t_d is twice the time taken to identify the SNR at one position and can be written as:

$$t_d \cong 2(t_c + t_o + B_\phi / \omega_\phi) \qquad (1.10)$$

The electric motor driving the antenna rotation must have the mechanical ability to change the direction of rotation rapidly. This is more difficult than running the antenna at a constant angular velocity. The output torque of the reversible motor and the weight of the antenna system must be commensurate.

There are many possible algorithms that can be used to reduce the time spent with the antenna pointing at adjacent positions. These include the simplex method and the gradient method. These are mathematical techniques used to locate the maximum or minimum value in a multidimensional surface (see Chapter 5). The simplest method is to perform an incremental scan of one beamwidth B_ϕ in the + and $-\phi$ directions. If the original position of the detected signal is ϕ_1, then the signal received at the two adjacent beam positions, $\phi_1 + B_\phi$, and $\phi_1 - B_\phi$ is compared to the signal strength of the beam received at ϕ_1. If either one is stronger than that received when the antenna is in the ϕ_1 direction, then the beam position is updated to the new direction of maximum signal strength. This dithering operation might be initiated if the signal level into the receiver begins to reduce, or periodically after a specified, preprogrammed period of time. The timing of this update is dependent on the specific application. For example, in the case of an airport radar, the velocity and direction of the incoming or outgoing aircraft must be taken into account when programming the dithering repeat time to ensure continuous tracking of the aircraft. It is also possible to retain the position history of the transmitter and use that to predict the next position.

1.3.2 Two-Dimensional Tracking

The algorithm for a 2D global search using sequential uniform sampling is given in Figure 1.5; the local search using the dither technique is given in Figure 1.6. In both cases, the SNR must be determined at each new direction of the antenna and the information stored for later comparison with other sampled values. The optimal direction is determined and then updated periodically as required.

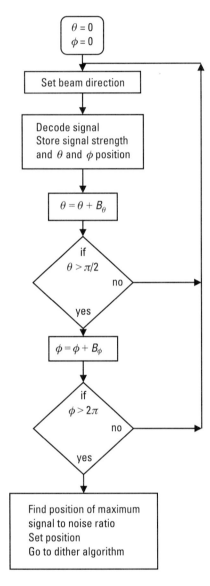

Figure 1.5 The algorithm for a two-dimensional search based on sequential uniform sampling.

A typical cellular telephone base station antenna is located on a tall pole (see Figure 1.7). The coverage of a single antenna is usually $2\pi/3$ radians (120°) in the ϕ plane and almost 90° in the θ plane. Consider the case of a

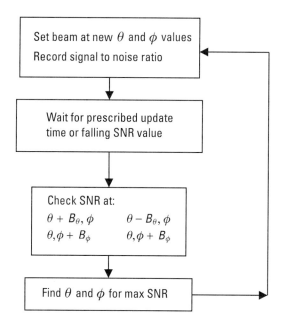

Figure 1.6 Two-dimensional dither algorithm to update the beam direction.

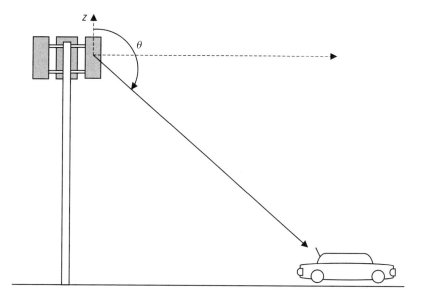

Figure 1.7 Base station tower for the cellular telephone network.

cellular telephone base station tracking antenna with control in two angular directions, the ϕ and θ directions. This antenna is required to track the radio signal from a moving vehicle on the ground below the tower.

The antenna has beamwidths B_θ and B_ϕ in the θ and ϕ directions, respectively. Assume that the maximum angular coverage of the antenna system is given by $\pi/2 < \theta < \pi$ and $0 < \phi < 2\pi/3$. Assuming equally spaced beam positions, the number of angular positions N_{tot} needed to cover the entire segment is given by:

$$N_{tot} = N_\phi \, N_\theta = \left(\frac{2\pi}{3B_\phi} \right)\left(\frac{\pi}{2B_\theta} \right) = \frac{\pi^2}{3B_\phi B_\theta} \tag{1.11}$$

The time required to complete the global scan, t_a, using sequential uniform sampling with one antenna beam, is given by (1.9); however, the time required to move the antenna is effectively zero for electronic control. The global scan time becomes:

$$t_a = N_{tot}(t_c + t_o) \tag{1.12}$$

1.4 Summary

This chapter outlined the major design requirements for a smart antenna. These issues relate to the radiation characteristics of the antenna and the timing requirements of the tracking system. Both of these characteristics are dependent on the application. Some descriptions were related solely to mechanically driven systems, because these are easier to understand. They are of course more generally applicable to all smart-antenna systems. More sophisticated approaches to target acquisition and tracking are available [15]. Some of these techniques rely on previous best-location information and use predictive techniques based on the calculation of the relative velocity vector between the source and receiver. Other methods use multiparameter optimization techniques such as the simplex and gradient methods, which rely on information from a number of beam scan positions to move rapidly to the optimal position. These optimization techniques are similar to those used for antenna design and are discussed in Chapter 5.

The design goals are identical for electronically controllable antennas, regardless of whether the application is radar, radio direction finding, intelligent communications systems, or radio mapping. The cellular telephone network is a major application area for smart-antenna technology.

The criteria outlined in this chapter will allow the reader to assess antenna designs presented in later chapters of this book, and to quantitatively evaluate the success of a particular design. In order to address these issues comparatively with other technologies of electronically controlled antennas, this book includes some descriptions of phased arrays in addition to discussions of switched parasitic antenna systems.

References

[1] Mortazowi, A., I. Itoh, and J. Harvey (eds.), *Active Antennas and Quasi-Optical Arrays*, New York: IEEE Press, 1999.

[2] Liberti, J. C., and T. S. Rappaport, *Smart Antennas for Wireless Communications—IS-95 and Third Generation CDMA Applications*, Upper Saddle River, NJ: Prentice Hall, 1999.

[3] Rappaport, T. S. (ed.), *Smart Antennas: Adaptive Arrays, Algorithms and Wireless Position Location,* Piscataway, NJ: IEEE Press, 1998.

[4] Okamoto, G. T., *Smart Antenna Systems and Wireless LANs*, Norwell, MA: Kluwer Academic Press, 1999.

[5] Kumar, A., *Fixed and Mobile Terminal Antennas*, Norwood, MA: Artech House, 1991.

[6] Hirasawa, K., and M. Haneishi (eds.), *Analysis, Design, and Measurement of Small and Low-Profile Antennas*, Norwood, MA: Artech House, 1992.

[7] Delisle, G., K. Hettak, and G. Lucas, "Intelligent Antennas for Future Wireless Communications," in *Modern Radio Science*, M. A. Stuchly (ed.), New York: Oxford University Press, 1999, pp. 125–150.

[8] Ramo, S., J. R. Whinnery, and T. van Duzer, *Fields and Waves in Communication Electronics*, 2nd ed., New York: John Wiley and Sons, 1984.

[9] Neff, H. P., *Basic Electromagnetic Fields*, 2nd ed., New York: Harper and Row, 1987.

[10] Smith, G. S., *An Introduction to Classical Electromagnetic Radiation*, Cambridge, U.K.: Cambridge University Press, 1997.

[11] Stutzman, W. L., and G. A. Thiele, *Antenna Theory and Design*, New York: John Wiley and Sons, 1981.

[12] Mailloux, R. S., *Phased Array Antenna Handbook,* Norwood, MA: Artech House, 1994.

[13] Lin, J., and T. Itoh, "Active Integrated Antennas," *IEEE Trans. Microwave Theory and Techniques*, Vol. 42, No. 12, December 1994, pp. 2186–2194.

[14] Stephens, D. R., *Phase-Locked Loops for Wireless Communications: Digital and Analog Implementations*, Norwell, MA: Kluwer Academic Press, 1998.

[15] Sabatini, S., and M. Tarantino, *Multifunction Array Radar: System Design and Analysis*, Norwood, MA: Artech House, 1994.

2

Wire Antenna Theory

2.1 Introduction

One approach to describing the performance of an antenna is based on directional performance [1, 2], presented in Chapter 1 in relation to smart antennas. Another approach, based on impedances, represents the antenna as a circuit or network. This description is equally valid, as the antenna will be incorporated into an RF circuit as part of the total communications system (e.g., see Figure 1.4).

Input impedance, beamwidth, and directivity are three basic parameters used to describe antenna performance. The input impedance, measured in ohms, is a complex number, and is the load impedance that the antenna presents to a connecting transmission line. The beamwidths B_θ and B_ϕ, measured in degrees, are defined in Section 1.2.1 as the angular width of the main beam between the −3 dB points in the direction of the maximum radiated signal strength θ_{max} and ϕ_{max}. The directivity of an antenna is a general description of the radiation pattern, given as the ratio of the radiated power in one direction relative to the overall radiated power [2]. It is commonly used to describe the maximum radiation in one direction only, but can also be used to describe the radiation in all directions. When the beam occupies the solid angle Ω_a defined by (1.3), then the directivity of this beam is $4\pi/\Omega_a$. Information about the front-to-back ratio, the location of nulls in the radiation pattern, and the relative intensity of the main side lobes can be obtained from the directivity. The maximum directivity of an antenna represents the maximum gain of the antenna if it is 100% efficient. Gain can be calculated from the product of the antenna radiation efficiency and the directivity, where the antenna radiation efficiency, e_r, is the ratio of the amount of power supplied to but not radiated by the antenna to the power supplied to the antenna. Gain can be expressed in dBi, the power relative to the total radiated power, assuming isotropic radiation.

The front-to-back ratio is normally defined in terms of the ratio of the radiated field strength in the direction of maximum radiation, say, $E(\theta_{max}, \phi_{max})$, relative to the radiation in the opposite direction, $E(180° + \theta_{max}, 180° + \phi_{max})$. If the antenna is located on a ground plane, then an alternative definition is required, as radiation below the ground plane is not possible. In this case, the field strength in the "back" direction is $E(\theta_{max}, 180° + \phi_{max})$.

If the field strength is greatest in the direction θ_{max}, the half-power beamwidth B_θ defined in Section 1.2.1 can be calculated by solving for θ_+ and θ_- in the equations:

$$E(\theta_+) = E(\theta_-) = E(\theta_{max}) / \sqrt{2} \qquad (2.1)$$

where θ_+ and θ_- are the directions closest to θ_{max}, where the field strength is equal to $1/\sqrt{2}$ of the maximum field strength, which is the −3-dB or half-power point. The beamwidth is calculated from $B_\theta = \theta_+ - \theta_-$.

The frequency variation of the input impedance and beamwidth are used to define the bandwidth of the antenna as the frequency range where these parameters conform to a specification. These parameters can be calculated from the current distribution in the antenna. In this chapter we outline the basic theory of conversion from a radio frequency (RF) electrical source to radiated energy in the form of an electromagnetic wave in an isolated wire. Antennas with more than one wire element are also considered.

The complex impedance of an antenna structure contains both a resistive component R_r and a reactive component X_a. The antenna impedance Z_a can be written as a lumped impedance:

$$Z_a = R_r + jX_a \qquad (2.2)$$

The resistive component R_r is called the radiation resistance, and results from the loss in energy due to radiation. Conduction and dielectric losses also contribute to this term, but in this discussion, their effect is assumed to be insignificant. If the conductive and dielectric elements of the antenna structure are relatively lossless, then this approximation is a good one. The input impedance of an isolated wire antenna can be calculated using analytical equations such as that derived using the induced EMF method in [2].

At RF, the feed line is a transmission line, and maximum energy is transferred to the antenna when the characteristic impedance of the transmission line Z_0 equals the total impedance of the antenna Z_a. A number of different terms from network theory are used to describe the effectiveness of the impedance match. For example, the reflection coefficient ρ at the junction between the end of the transmission line and the input impedance of the antenna is given by the expression:

$$\rho = \frac{Z_a - Z_0}{Z_a + Z_0} \tag{2.3}$$

The voltage standing wave ratio (VSWR) is a measure of the strength of the reflection, and is given in terms of ρ:

$$\text{VSWR} = \frac{1 + |\rho|}{1 - |\rho|} \tag{2.4}$$

The magnitude of the scattering parameter S_{11}, usually expressed in dB [3], for the two-port network consisting of the transmission line and the antenna, is given by:

$$S_{11} = 20 \log_{10} |\rho| \tag{2.5}$$

The phase of S_{11} or ρ can also be calculated; however, the phase will not be discussed in this book, and so any reference to S_{11} relates to (2.5). This parameter is useful in link budget calculations where gains and losses over the entire link from RF transmitter electronics to RF receiver electronics are expressed in dB. The overall signal strength into the receiver can be estimated by summing the dB values for each loss and gain term in the link, assuming a fixed RF transmitter power. The impedance bandwidth is often defined as the frequency range where $S_{11} < -10$ dB.

In order to use a standard measure in this book, it will always be assumed that the transmission line has a characteristic impedance of $Z_0 = 50\Omega$. This means that an antenna is perfectly matched to the transmission line when the antenna impedance Z_a is exactly 50Ω. In this case:

$$\rho = 0, \qquad \text{VSWR} = 1, \qquad \text{and } S_{11} = -\infty \, \text{dB}$$

As Z_a is a function of frequency, so S_{11} is also frequency dependent. In this book, the resonant frequency f_0 of the antenna is defined as the frequency where the magnitude of the reflection coefficient is a minimum. For a perfect match between the antenna and the transmission line, the antenna input impedance must be purely real, that is, the reactance of the antenna X_a must be zero. While it is possible to use impedance-matching networks to eliminate the reactive component of an antenna and also to change the effective resistance of an antenna, such networks can have significant losses and are often highly frequency dependent [3]. These are undesirable consequences, and a more robust antenna design is achieved when a close match to a transmission line impedance is achieved by varying the physical characteristics of the antenna alone. This is the approach used in this book.

2.2 Basic Radiation Theory

A fundamental law in physics is that a radiated electromagnetic field is generated by accelerated charge [4]. This acceleration occurs at a number of possible locations in the antenna structure and can also occur in the feed structure. Radiation emanates from:

- The junction between the feed transmission line and the antenna;
- Bends in the antenna wires;
- The ends of the wires in the antenna;
- Changes in thickness of the wire;
- Discontinuities in the transmission-line feed structure.

In order to derive the radiation characteristics of an antenna, it is necessary to calculate the currents in all of the elements that form the antenna and to project the fields from these currents into the volume around the antenna. This calculation requires the application of Maxwell's equations and is usually performed analytically or numerically using the magnetic vector potential A [1, 2].

Assuming a three-dimensional current distribution J, usually sinusoidal, with angular frequency ω, located in a conductor with finite volume V', (see Figure 2.1), then A is given by the equation:

$$A(r,\omega) = \frac{\mu}{4\pi} \int_{V'} \frac{J(r',\omega)e^{-jkR}}{R} dV' \qquad (2.6)$$

where $R = |r - r'|$ and μ is the scalar magnetic permeability, $j = \sqrt{-1}$, $k = 2\pi/\lambda_0$ is the wave number, and λ_0 is the wavelength of the radiation in free space. The position vector r' is defined in Figure 2.1. It locates the current distribution in elemental volume dV', which is part of the total volume V' containing the current J. The position vector r locates the observation point some distance away from both the origin and the antenna volume.

Equation (2.6) assumes that the magnetic vector potential is defined by the equation:

$$B = \nabla \times A \qquad (2.7)$$

where B is the magnetic flux density vector.

Assuming that the receiver lies in the far field of the current distribution (i.e., $|r| >> |r'|$), then the two vectors r and $r\text{-}r'$ can be considered parallel. Then [2],

$$A(r,\omega) = \frac{\mu}{4\pi} \frac{e^{-jkr}}{r} \int_{V'} J(r',\omega)e^{jkr'\cos\xi} dV' \qquad (2.8)$$

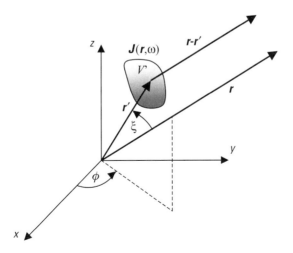

Figure 2.1 Coordinate system showing a 3D conductor having volume V' carrying volume current $J(r,\omega)$ detected at a distant point from the origin at vector r.

where ξ is the angle between the two vectors r and r', as shown in Figure 2.1. This general expression can be used with (2.7) to calculate the field radiated from a three-dimensional current distribution J.

2.3   Radiation from a Wire Element

If the current is confined to a single straight wire lying along the z axis (see Figure 2.2), $|J| = J_z$ and the electric field in the far field has a θ component only, that is,

$$A_z = \frac{\mu}{4\pi} \frac{e^{-jkr}}{r} \int_{V'} J_z(r') e^{jkr' \cos \xi} \, dV' \tag{2.9}$$

If the wire is very thin, the current J_z can be written as:

$$J_z(z') = I_z(z')\delta(x')\delta(y') \tag{2.10}$$

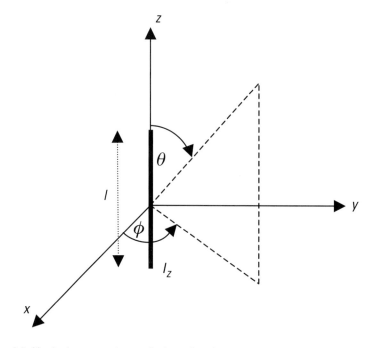

Figure 2.2 Vertical current element in the z direction.

where δ is the unit impulse function or the dirac delta function [1], defined in the following way:

$$\delta(x') = \begin{cases} 1 \text{ for } x' = 1 \\ 0 \text{ for } x' \neq 1 \end{cases} \quad (2.11)$$

Equation (2.11) indicates that the current distribution in the x' and y' directions is infinitely thin.

Substituting (2.10) into (2.8) results in:

$$A_z = \frac{\mu e^{-jkr}}{4\pi r} \int_{-\infty}^{\infty} \delta(x')dx' \int_{-\infty}^{\infty} \delta(y')dy' \int_{-\infty}^{\infty} I_z(z')e^{jkz'\cos\theta}dz' \quad (2.12)$$

and these three 1D integrals can be calculated independently.

From (2.11) it follows that:

$$\int_{-\infty}^{\infty} \delta(x')dx' = \int_{-\infty}^{\infty} \delta(y')dy' = 1 \quad (2.13)$$

and (2.12) becomes:

$$A_z = \frac{\mu e^{-jkr}}{4\pi r} \int_{-\infty}^{\infty} I_z(z')e^{jkz'\cos\theta}dz' \quad (2.14)$$

Using (2.7) and converting to spherical polar coordinates, the only nonzero magnetic field component is:

$$B_\phi = jkA_z \sin\theta \quad (2.15)$$

From the time harmonic form of Maxwell's equations $j\omega\varepsilon E = \nabla \times H$, the only nonzero electric field component from (2.15) is:

$$E_\theta = j\omega A_z \sin\theta \quad (2.16)$$

The electric field E_θ defines the far-field radiation pattern of an antenna and can be written directly in terms of the current distribution I_z by combining (2.16) and (2.14), that is,

$$E_\theta = \frac{j\omega\mu e^{-jkr} \sin\theta}{4\pi r} \int_{-\infty}^{\infty} I_z(z')e^{jkz'\cos\theta} dz' \qquad (2.17)$$

The integral term in (2.14) and (2.17) is very similar to the Fourier transform of the current distribution [1, 2]. The Fourier transform $F(x)$ of the function $g(s)$ is defined by the expression:

$$F(x) = \frac{1}{2\pi} \int_{-\infty}^{\infty} g(s)e^{jxs} ds \qquad (2.18)$$

and the inverse Fourier transform can be written as:

$$g(s) = \int_{-\infty}^{\infty} F(x)e^{-jsx} dx \qquad (2.19)$$

Comparing (2.14) and (2.18), the required substitutions are:

$$x = \cos\theta \quad \text{and} \quad s = kz' \qquad (2.20)$$

The distances must be normalized by the wave number, and the radiation pattern is given in terms of uniformly distributed $\cos\theta$. If the magnitude of the radiation pattern is all that is required, then the constants in front of the integral sign in (2.17) and (2.18) can be ignored. The phase term, however, given by the exponential term, must be retained in cases when there is more than one element in the array.

The radiation characteristics can be determined for a number of cases in which the current distribution is z-directed.

2.3.1 Hertzian Dipole

An infinitesimally short current element oriented in the z direction carrying uniform current I_0 can be represented by the equation:

$$I_z = I_0 \Delta z' \tag{2.21}$$

Substituting (2.21) into (2.17) results in:

$$E_\theta = \frac{j\omega\mu\Delta z' I_0 e^{-jkr}}{4\pi r} \sin\theta \tag{2.22}$$

This means that the radiation pattern is independent of the ϕ direction and has $\sin\theta$ dependence only. The radiation pattern in the plane parallel to the electric field, in this case the θ pattern, is commonly referred to as the E plane radiation pattern, and that parallel to the magnetic field is referred to as the H plane radiation pattern.

The radiated field has only one component of the E field and only one component of the H field. This type of radiation field is referred to as being linearly polarized. A single straight wire radiator generates linearly polarized radiation. In practice, currents in the transmission line feed cable and some imperfections in the mechanical construction of the antenna can result in additional components in the radiation field. The copolar radiation pattern consists of the dominant polarization component in the radiated field. For straight wire antennas, the radiation is linearly polarized and so the dominant polarization component is parallel to the wire. The cross-polar radiation component consists of the orthogonal polarization components in the far field. If a straight wire is vertically polarized, the orthogonal radiation component is the component of the radiation pattern that is horizontally polarized.

The current element described by (2.21) cannot be constructed in practice, but is used theoretically as the fundamental radiation element called a Hertzian dipole. The far-field radiation from any antenna structure can be determined by summing the radiation from all infinitesimally small current elements in the structure, taking note of the current magnitude, current phase, orientation, and position.

Using (2.1) with $\theta_{max} = 90°$, this gives $B_\theta = 90°$ for the Hertzian dipole. As there is no ϕ dependence, B_ϕ is the maximum value possible, 360°.

2.3.2 Short Dipole with Uniform Current

As a second example, consider a conducting element of length l, oriented in the z direction, carrying uniform current I_0. This current distribution can be written as:

$$I_z(z') = \begin{cases} I_0 & -l/2 \le z' \le l/2 \\ 0 & |z'| > l/2 \end{cases} \qquad (2.23)$$

Substituting this expression into (2.17) gives:

$$E_\theta = \frac{j\omega\mu\, l I_0 e^{-jkr}}{4\pi r} \sin\theta \, \frac{\sin[(kl/2)\cos\theta]}{(kl/2)\cos\theta} \qquad (2.24)$$

Using (2.24), it can be shown that the beamwidth B_θ is given by the approximate relationship:

$$\sin B_\theta \cong \frac{1}{\sqrt{2}} \frac{1}{[1-(kl)^2/12]} \quad \text{for } kl < 0.5 \qquad (2.25)$$

The term $\sin[(kl/2)\cos\theta]/(kl/2)\cos\theta$ in (2.24) is a sinc function, and can be identified with the single slit interference expression well known in optics [5]. Note that in optics the angle θ is measured from the line perpendicular to the slit, whereas in these expressions θ is measured from the axis parallel to the current distribution. Thus, the cosine expression within the square brackets is replaced by a sine expression. In optics, the Fourier transform of the aperture distribution is the far-field angular distribution [5]. The integral used to calculate the radiation pattern for the Hertzian dipole represents radiation in all directions. This is identical to a point aperture source in optics multiplied by the $\sin\theta$ term. This analogy can be used for all wire current elements and aperture antennas, where the electric field distribution across an aperture can be related to a fictitious current distribution defined by the field distribution [2, 6].

2.3.3 Dipole with Sinusoidal Current Distribution

The Hertzian dipole and the current element of finite length carrying uniform current are not real. This is because the current at the ends of the conducting element is not zero. A practical antenna must have zero current at the ends of the conducting element.

A more realistic description of a radiation source is a dipole with a sinusoidal current distribution along its length and $I_z(z') = 0$ at $z' = l/2$. The current can be expressed by the equation:

$$I_z(z') = \begin{cases} I_0 \sin k(l/2 - z') & 0 < z' < l/2 \\ I_0 \sin k(l/2 + z') & -l/2 < z' < 0 \\ 0 & |z'| > l/2 \end{cases} \quad (2.26)$$

Substituting (2.26) into (2.17) gives:

$$E_\theta = \frac{j\omega\mu 2lI_0 e^{-jkr}}{4\pi r} \left[\frac{\cos\left(\dfrac{kl}{2}\cos\theta\right) - \cos\dfrac{kl}{2}}{\sin\theta} \right] \quad (2.27)$$

When $l = \lambda_0/2$, $kl = \pi/2$ and the radiation pattern is given by:

$$E_\theta = \frac{j\omega\mu I_0 e^{-jkr}}{2\pi rk} \left[\frac{\cos\left(\dfrac{\pi}{2}\cos\theta\right)}{\sin\theta} \right] \quad (2.28)$$

This is a common and practical wire antenna element, referred to as the half-wave dipole. The −3-dB beamwidth of this antenna $B_\theta \cong 78°$.

The radiation patterns in the θ plane for these three antennas (Hertzian dipole, straight wire with $l = 0.3\lambda_0$ and uniform current, and the half-wave dipole) are given in Figure 2.3. Each of the three radiation patterns has been normalized to have a maximum gain of unity in the principal beam direction θ_{max}. For all three antennas, $\theta_{max} = 90°$.

2.3.4 Dipole with Linear Current Distribution

As a final example, a short wave dipole with length less than $\lambda_0/2$, but with zero current at the ends of the elements is addressed. The current distribution can be approximated by a linear function:

$$I_z(z') = \begin{cases} I_0(l/2 - z') & 0 < z' < l/2 \\ I_0(l/2 + z') & -l/2 < z' < 0 \\ 0 & |z'| > l/2 \end{cases} \quad (2.29)$$

The radiation pattern for this current distribution is given by the expression:

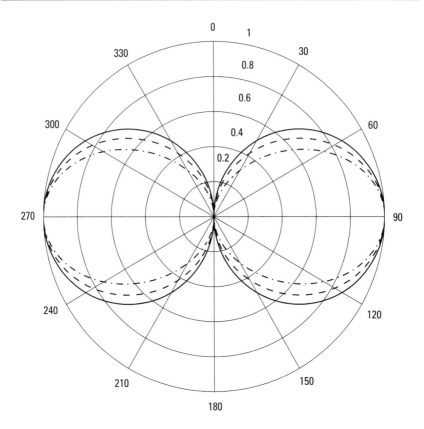

Figure 2.3 E plane radiation patterns (θ plane) for three basic wire dipoles: '____' Hertzian dipole, '_ _ _' half-wave dipole, and '_ . _' short dipole ($l = 0.3\lambda_0$) with uniform current.

$$E_\theta = \frac{j\omega\mu\, lI_0 e^{-jkr}}{4\pi r} \sin\theta \, \frac{\sin^2\left(\dfrac{kl}{4}\cos\theta\right)}{\left(\dfrac{kl}{4}\right)^2 \cos^2\theta} \tag{2.30}$$

The radiation patterns for dipoles with $l \leq \lambda_0/2$ length given by (2.22), (2.24), (2.28), and (2.30) are similar in shape. They differ in their efficiency in converting the input current to a radiated field, and this can be described in part by the input impedance of the antenna.

2.3.5 Impedance of Wire Antennas

It can be shown that R_r is strongly dependent on the total length l of the dipole antenna [1, 2, 6, 7]. For example:

Hertzian dipole: $\qquad R_r = 20(kl/2)^2$

Short dipole: $\qquad\qquad R_r = 5(kl/2)^2$ $\qquad\qquad\qquad$ (2.31)

Half-wave dipole: $\qquad R_r = 73.08$ ohms

The reactance of a thin wire dipole antenna X_a is a measure of the energy stored in the nonradiated fields. It is approximately equal to the reactance of a parallel wire transmission line that is terminated in an open circuit [2, 7]. Assume that a parallel wire transmission line has been opened out at a length $l/2$ from its open-circuit end, and the two wires bent so that they are collinear and perpendicular to the direction of the transmission line. This transmission line assumption gives:

$$X_a \cong -Z_0 \cot kl \qquad\qquad (2.32)$$

When $kl = \pi/2, 3\pi/2, \ldots$, then $X_a = 0$ and Z_0 is real. A half-wave dipole is close to the lowest resonant frequency, that is, when the length of the dipole is approximately $\lambda_0/2$, the antenna impedance is almost purely real.

The input impedance of a dipole antenna can be altered by changing the position of the feed point. If Z_a is the impedance when the antenna is fed at its center, the impedance of an offset feed point Z_{as} is approximately given by [2]:

$$Z_{as} \cong \frac{Z_a}{\cos^2 k\Delta l} \qquad\qquad (2.33)$$

where Δl is the distance from the center point. Figure 2.4 shows the effect on the reflection coefficient of a dipole antenna resulting from off-center feed positions at 48%, 42%, and 36% of the length of the element. The data was obtained using the method of moments (MOM)–based numerical electromagnetics code (NEC) [8]. While there is no change in the resonant frequency, the change in the 10-dB bandwidth is clear. The center-fed dipole has the largest bandwidth.

An extension to this idea is to feed the dipole symmetrically at two points spaced from the center of a continuous dipole element. This configuration, referred to as the "tee-match" configuration, is shown in Figure 2.5

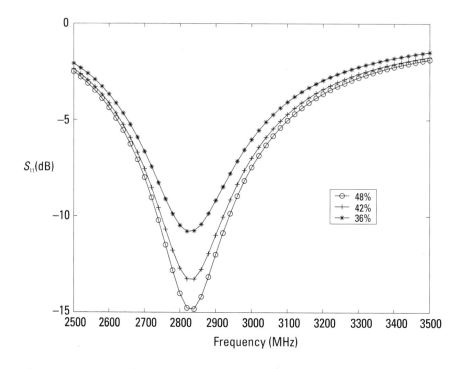

Figure 2.4 Frequency variation in the S_{11} (dB) of a half-wave dipole for various feed positions. The dipole length is 0.05m and the wire radius is 0.001m.

where the inner section of the feed line forms a short-circuited parallel transmission line of length $2\Delta l$. The input impedance of this antenna Z_{as} is given by [9]:

$$Z_{as} \cong \frac{(1+a_r)^2 Z_a Z_f}{Z_f + (1+a_r)^2 Z_a} \tag{2.34}$$

where $Z_f = jZ_l \tan k_l \Delta l$ is the impedance of the parallel wire transmission line segment having characteristic impedance Z_l, k is the wave number for the transmission line, and a_r is the ratio of the diameters of the two wires forming the transmission line.

A number of analytical formulations have been derived to calculate the input impedance of a dipole of arbitrary length and wire diameter [10]. Numerical methods are commonly used to calculate the input impedance of

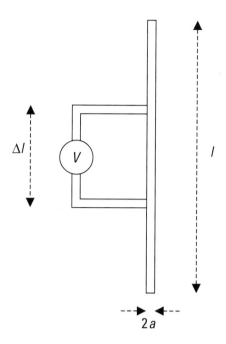

Figure 2.5 Tee-match configuration for a dipole antenna with wire radius *a*.

wire antennas [8]. In particular, NEC is an efficient method for solving these problems. As an example, Figure 2.6 shows the effect of varying the wire radius a on the S_{11} of a center-fed dipole with length $\lambda_0/2$ calculated using NEC [8]. S_{11} is plotted as a function of frequency. The impedance bandwidth of the antenna increases with increasing wire radius a while the resonant frequency f_0 of the antenna decreases slightly.

In a practical situation it may be desirable to use antenna elements that do not have a circular cross section. In these cases, an effective wire radius can be used for the calculation of radiation performance. This equivalence covers conductors with a rectangular cross section, elliptical cross section, and L-shaped (sometimes referred to as corner) conductors [2, p. 456].

2.4 Mutual Coupling Between Wire Elements

When a wire antenna is placed close to a dielectric body, displacement currents may be induced in the dielectric. Similarly, if a wire antenna is placed close to a conducting body, conduction currents may be induced to flow in

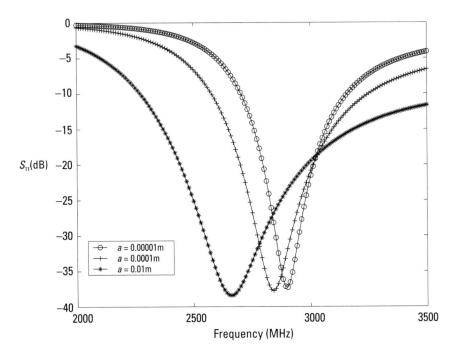

Figure 2.6 Variation in the S_{11} (dB) of a half-wave dipole (length = 0.05m) as a function of frequency for three different radius values a.

that body. This has the effect of changing the current distribution in the original antenna and so affects its input impedance. The currents in the neighboring body contribute to the total radiation pattern of the antenna. In switched parasitic antennas, these induced currents are of significant interest. When the size of these adjacent elements approaches a resonant size, then the induced currents are maximized and the effect on the total antenna system may be significant. Such conductors are called parasitic elements, and need not be the same form as the driven elements. For example, a parasitic slot in a ground plane has been used to modify the radiation characteristics of an array of wire monopole elements [11]. This induction effect is often referred to in terms of the mutual impedance, or as an image if the object is large.

Consider the case of two conducting wires lying close to each other and driven with voltages V_1 and V_2. The currents in the two elements I_1 and I_2 can be calculated by solving the network equation:

$$V_1 = Z_{11}I_1 + Z_{12}I_2$$
$$V_2 = Z_{21}I_1 + Z_{22}I_2 \qquad (2.35)$$

where Z_{11} and Z_{22} are defined as the impedance looking into port 1 or 2 when all other ports in the network are open circuited [2]. In a p element array:

$$Z_{nn} = \left. \frac{V_n}{I_n} \right|_{I_m = 0 .. p, m \neq n} \tag{2.36}$$

where $m = 1, 2, ...p$, and $n = 1, 2, ...p$. Similarly, Z_{12} and Z_{21} are the mutual impedances between ports 1 and 2 or 2 and 1. In a network with more than two ports, the mutual impedance between any two of the ports is calculated with the remaining ports open circuited:

$$Z_{mn} = \left. \frac{V_m}{I_n} \right|_{I_m = 0 .. p, m \neq n} \tag{2.37}$$

If the wire elements are resonant at f_0 and have feed points close to the center of the wire, an open-circuited feed results in two nonresonant wires, and little current is induced in the two short wires. For an array of resonant length wires, the calculation of Z_{11} and Z_{22} are simplified by ignoring other elements of the array, and so these are approximately equal to the self-impedances of the individual elements in isolation. Similarly, Z_{12} and Z_{21} can be calculated in isolation and so are the mutual impedances between the two isolated elements. These approximations are valid when there is minimal induced current in the array elements with their feed position open circuited.

In the case of many such elements, (2.35) can be expanded to include the influence of every element in the array on every other element in the array. This can be written as the impedance matrix equation from network analysis:

$$\begin{bmatrix} V_1 \\ V_2 \\ . \\ . \\ V_n \end{bmatrix} = \begin{bmatrix} Z_{11} & Z_{12} & . & . & Z_{1n} \\ Z_{21} & Z_{22} & . & . & Z_{2n} \\ . & & . & . & . \\ . & & . & . & . \\ Z_{n1} & Z_{n2} & . & . & Z_{nn} \end{bmatrix} \begin{bmatrix} I_1 \\ I_2 \\ . \\ . \\ I_n \end{bmatrix} \tag{2.38}$$

If all elements have the same length and each is driven at the center, then $Z_{11} = Z_{22} = Z_{nn} = Z_a$, where Z_a is the input impedance of the dipole in isolation. Methods of calculating this impedance were given in Section 2.3. When the voltages V_1, V_2, ..., V_n on the left-hand side of (2.38) are known, the impedance matrix elements can be calculated independently, and so the currents I_1, I_2, ..., I_n in each of the n elements can be calculated.

The far-field radiation pattern E_{tot} of the array consisting of n elements is given by the sum of the radiation patterns of the elements individually, where the currents are represented by complex numbers and the phase differences in all directions are taken into account. The radiation pattern E_{tot} can be written in the form [1, 2]:

$$E_{tot} = \sum_{m=1}^{n} E(I_m) \tag{2.39}$$

where $E(I_m)$ is the radiation pattern for the mth element carrying current I_m. The phase and magnitude of the voltages V_m in (2.38) can be used to control the direction and beamwidth of the antenna array. This is discussed in detail in Section 2.6.

Returning to the two-element array described using (2.35), if one element (say, element number 2) is not driven and is short-circuited at its feed position, then $V_2 = 0$ and (2.35) can be written as [12]:

$$V_1 = Z_{11}I_1 + Z_{12}I_2$$
$$0 = Z_{21}I_1 + Z_{22}I_2 \tag{2.40}$$

The radiation pattern is still given by (2.39), where $n = 2$. The calculation of the radiation characteristics of both phased arrays and parasitic antennas requires solution of the impedance matrix (2.38), which includes the self-impedances (the on-diagonal elements in the matrix and previously described by the symbol Z_a) and the mutual impedances (the off-diagonal elements in the matrix). The calculation of the mutual impedances is an important part of the derivation of the antenna characteristics. For thin wire antennas, the calculation can be done analytically [1, 2, 10] or numerically using the MOM code [8].

Figure 2.7 illustrates two configurations of two parallel dipoles located in close proximity to each other. The elements are driven at the center point by two different voltages, V_1 and V_2. The first arrangement is called the side-by-side configuration [Figure 2.7(a)], and the second is called the collinear configuration [Figure 2.7(b)] [2]. These two cases are discussed separately.

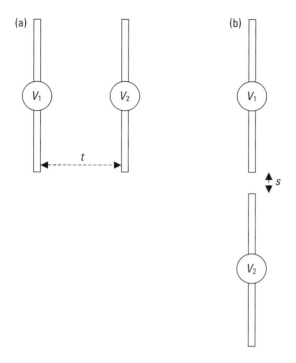

Figure 2.7 Two parallel dipoles independently driven with voltages V_1 and V_2: (a) side-by-side configuration, and (b) collinear configuration.

2.4.1 Side-by-Side Configuration

Consider two parallel wire antenna elements of approximate resonant length [Figure 2.7(a)] and driven by voltages V_1 and V_2, respectively. Figure 2.8 shows the variation in the mutual impedance $Z_{12} = R_{12} + jX_{12}$, as a function of the separation distance t between two half-wave dipoles. These results were calculated using the analytical expressions for the induced EMF method [2].

Consider the impedance matrix for a four-element linear array of half-wave dipoles, in which the element spacing $t = 0.2\lambda_0$. Figure 2.8 can be used to obtain the mutual impedance elements of the impedance matrix, and these are given in Table 2.1.

If the voltage applied to each element is 1 volt, then the currents in each of the elements calculated from (2.38) with $n = 4$ are:

$$I_1 = 8.3 - j1.9 \text{ mA}$$
$$I_2 = 5.7 + j3.7 \text{ mA}$$

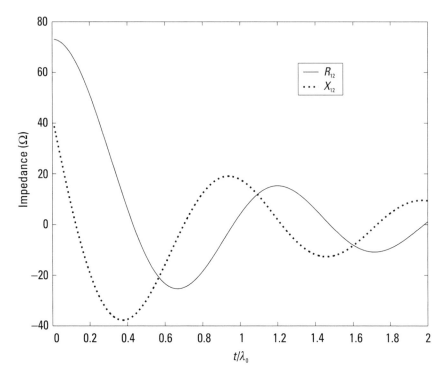

Figure 2.8 Variation in the real and imaginary components of the mutual impedance Z_{12} (Ω), for two parallel half-wave dipole elements separated by distance t/λ_0, in the side-by-side configuration.

$$I_3 = 5.7 + j3.7 \text{ mA}$$
$$I_4 = 8.3 - j1.9 \text{ mA}$$

Figure 2.8 can be used to calculate the mutual impedance for both adjacent and nonadjacent elements in the side-by-side array. This is a direct consequence of the principle of superposition.

Figure 2.9 illustrates the change in S_{11} for three different spacings of a two-element dipole array (side-by-side configuration) with one driven element and one short-circuited parasitic element of the same length. Results are compared with data for a single antenna in free space, which is equivalent to an infinite spacing. The effect of decreasing the spacing is to decrease the bandwidth of the antenna and to shift the resonant frequency.

Table 2.1
Impedance Matrix Elements for a Four-Element Dipole Array with an Interelement Spacing of $t = 0.2\lambda_0$*

Displacement (λ_0)	Matrix Elements	Impedance Value (Ω)
	$Z_{11} = Z_{22} = Z_{33} = Z_{44}$	$73.0 - j40.1$
0.2	$Z_{12} = Z_{21} = Z_{23} = Z_{32} = Z_{34} = Z_{43}$	$51.4 - j19.2$
0.4	$Z_{13} = Z_{31} = Z_{24} = Z_{42}$	$6.2 - j37.4$
0.6	$Z_{14} = Z_{41}$	$-23.3 - j15.9$

* The induced EMF method was used to calculate the impedance values [2].

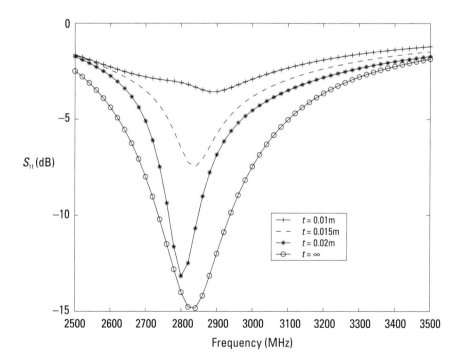

Figure 2.9 S_{11} (dB) as a function of frequency for a two-dipole array (one active and the other parasitic) in the side-by-side configuration. The separation distance t between the two elements ranges from infinity (i.e., a dipole in free space) to 0.01m. The dipole length is 0.05m with a radius of 0.0001m.

The effects of mutual coupling have significant consequences for antenna arrays. If the elements are to act independently of each other in the array, then the distance between the elements must be large enough to ensure that the mutual coupling is insignificant. For an array of parallel half-wave dipoles, Figure 2.8 suggests that the separation distance must be greater than $2\lambda_0$. In both switched parasitic antennas and Yagi-Uda antennas [1, 2], parasitic elements are used to alter the radiation direction and so must be located close to the current-carrying element.

2.4.2   Collinear Configuration

In a similar manner, the mutual coupling between two dipole antennas located along the same line—the collinear configuration [see Figure 2.7(b)]—can be investigated. The network equations (2.35) and (2.37) are used to calculate the current in each element; however, unlike the side-by-side configuration, the far-field radiation-pattern variation occurs in the E plane. From symmetry, the H plane radiation pattern of this configuration remains a circle centered on the line joining the antennas.

The mutual impedance between the two collinear elements is less significant than that for the side-by-side case. Impedance values $Z_{12} = R_{12} + jX_{12}$, obtained using the induced EMF method [2] for two half-wave dipoles, are shown in Figure 2.10. The effect of a half-wavelength collinear parasitic element on the S_{11} of the driven element is given in Figure 2.11 for a number of separation distances s.

2.5   Bandwidth of Wire Antennas

The bandwidth of an antenna is the frequency range over which the radiation direction (including the beamwidth) and impedance characteristics are maintained to within a predefined level. This broad definition allows the user to define what is important as far as the antenna performance is concerned, and can be different for different applications. The user might require that the front-to-back ratio or the level of the maximum sidelobes of the antenna conform to a specification for all frequencies within the band. One important parameter is the reflection coefficient S_{11}, outlined in Section 2.1. This quantifies how much power from the transmitter is reflected from the junction between the transmission line and the antenna, and so is related to the total power radiated from the antenna. The bandwidth can also be derived from the radiation pattern of the antenna. The impedance bandwidth,

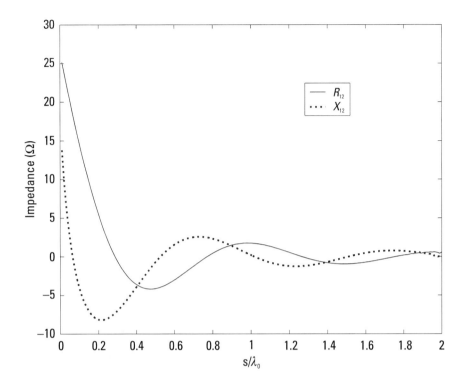

Figure 2.10 Variation in the real and imaginary components of the mutual impedance for two collinear half-wave dipole elements as a function of the separation distance *s*.

defined in terms of S_{11}, is a common definition used by antenna engineers, as it is simple to define and measure.

One commonly accepted value for defining the bandwidth is to choose the frequency range where $S_{11} < -10$ dB. This limit implies that the antenna performs satisfactorily only if the magnitude of the reflection coefficient $|\rho| < 0.32$, or less than 32% of the input voltage, or approximately 10% of the power, is reflected.

The S_{11} of a wire dipole antenna of length $l = 0.05$m with a feed cable impedance of 50Ω is given in Figure 2.6 for three different values of the wire radius *a*. The data was calculated using NEC [8] for a range of frequencies. The resonant frequency decreases and the impedance bandwidth increases as the wire radius increases. For $a = 10^{-5}$m, the resonant frequency is 2.9 GHz and the -10-dB bandwidth is approximately 460 MHz.

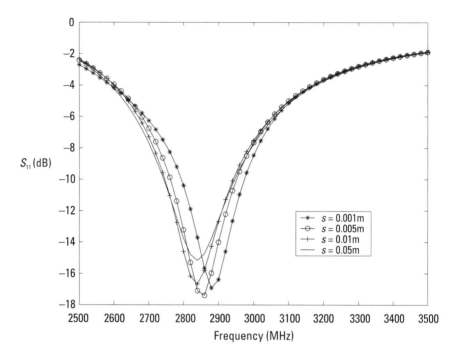

Figure 2.11 S_{11} (dB) plotted as a function of frequency for a driven dipole antenna with one parasitic element in the collinear configuration. The separation distance s between the two elements ranges from 0.001m to 0.05m. The dipole length is 0.05m with a radius of 0.0001m.

The bandwidth can also be described as a percentage variation relative to the center frequency. For this antenna, the percentage bandwidth is 16%. For some applications, a wider bandwidth will be required and a number of techniques can be used to increase the bandwidth. These techniques include an increase in the wire radius a, the use of parasitic antennas located close to the driven antenna, lumped-impedance loading, and various impedance-matching strategies. Some of these strategies are discussed later in this chapter.

2.6   Wire Antennas Above a Finite Ground Plane

In Section 2.4, it was shown that a conducting wire located close to a radiating element has an induced current that affects the current in the driven element. The consequence is a change in the antenna impedance and a change in the antenna directional characteristics. The same is true if a conducting or insulating body (assuming its relative permittivity ε_r is different from that of free space) is brought relatively close to a radiating element.

A simple, yet practical, case is the situation where a current-carrying conductor is located in the vicinity of a perfectly conducting ground plane of infinite extent. The current distribution in the ground plane is such that the total field can be determined by replacing the ground plane with an additional current element located at the mirror image position behind the plane. This imaginary element, called an image, carries a current with magnitude identical to the source element, with the component perpendicular to the surface being in phase with the source, and with the component parallel to the surface 180° out of phase with the current in the source. This is illustrated in Figure 2.12. The input impedance and the radiation pattern of the driven element can be calculated using the mutual impedance equations given in the previous section. The radiation pattern cannot extend below the infinite ground plane, and so only the upper hemisphere is relevant to the directional characteristics of the antenna.

If an end-driven wire element is located perpendicular to and immediately above the ground plane, the image current element combines with the real current element to form an effective isolated dipole of twice the length. If the length of the monopole element is one-quarter wavelength, then this antenna configuration is referred to as a quarter-wave monopole. If the ground

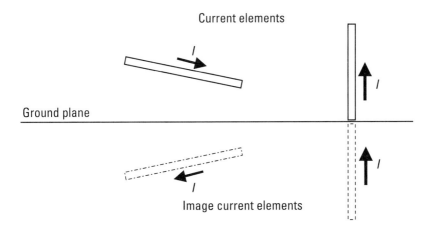

Figure 2.12 Image-current element representation of a current element in close proximity to a perfectly conducting ground plane of infinite extent.

plane is of infinite extent, the radiation pattern in the upper hemisphere is identical to that of an equivalent dipole, but the radiation resistance is one-half that of the half-wavelength dipole, as only the upper hemisphere is illuminated [1, 2]. This configuration is shown in Figure 2.13 where the element is fed from underneath the ground plane with a coaxial cable. This arrangement can be used for both phased arrays and parasitic antennas, where the additional elements are also monopoles. One advantage of the monopole structure is that the feed cables for the additional elements can be located beneath the ground plane and so do not contribute to the radiation pattern. In dipole antennas, the feed cables can adversely affect the radiation pattern, in particular the cross-polar component.

The characteristics of an isolated monopole on a ground plane are significantly affected if the ground plane is not of infinite extent. Consider the case when a coaxial cable feeds the monopole without a ground plane (see Figure 2.14). In this case, the outer surface of the shield of the coaxial cable carries current induced by the current in the monopole. Both the input impedance and the radiation pattern are influenced by changes in this current along the entire length of the feed cable. It has been found experimentally that placing a hand on the cable some distance from the antenna causes a change in the impedance and/or the radiated field strength.

One solution to this problem is to include a reflection point on the outer part of the coaxial cable, at one-quarter wavelength away from the feed. This can be achieved either through an external conducting sleeve with a length approximately $\lambda_0/4$ [2, 9], a set of radial wires of length $\lambda_0/4$, or a ferrite bead located approximately $\lambda_0/4$ from the end of the coaxial cable [13].

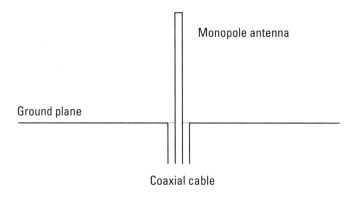

Figure 2.13 Monopole element on a ground plane.

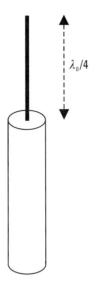

Figure 2.14 Monopole driven by a coaxial cable without a ground plane.

These antennas are illustrated in Figure 2.15. Both active and parasitic elements in an array can be configured in this way.

Returning to the case of a monopole above a ground plane of limited extent, an approximate solution for the radiation pattern can be obtained using an integral method such as that developed by Richmond [14, 15]. Figure 2.16 is a schematic diagram of a monopole centered on a circular ground plane of radius t. It is assumed that the ground plane is perfectly conducting and infinitely thin. Currents on the coaxial feed cable are not taken into account. Figure 2.17 illustrates the effect of t on the E plane radiation pattern using data taken from [15]. The data has been normalized to a maximum gain of 0 dB for each plot, and the $\sin\theta$ dependence of a short current element using (2.22) has been plotted for comparison. For $t < 0.3\lambda_0$, the radiation pattern is almost identical to that of an electrically short current element. In the range $t > 0.3\lambda_0$, the radiation has maximum gain in the direction $0° < \theta < 90°$ and the direction of maximum gain in the θ plane is dependent on t.

The radiation resistance of the monopole increases significantly as t increases, so the radiation efficiency of the antenna structure increases proportionately. The ground plane is resonant at a number of different radii and the number of lobes in the radiation pattern corresponds to these resonances.

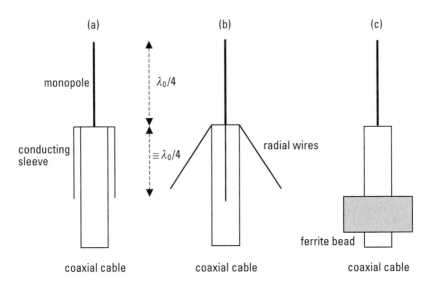

Figure 2.15 Monopole element at the end of a coaxial feed cable: (a) quarter-wavelength sleeve, (b) quarter-wavelength radial wires, and (c) ferrite bead at one-quarter wavelength.

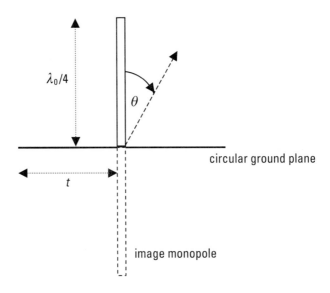

Figure 2.16 Schematic diagram of a monopole wire element centered on a circular ground plane of radius t. The dashed line represents the image of the monopole present when the ground plane is of infinite extent.

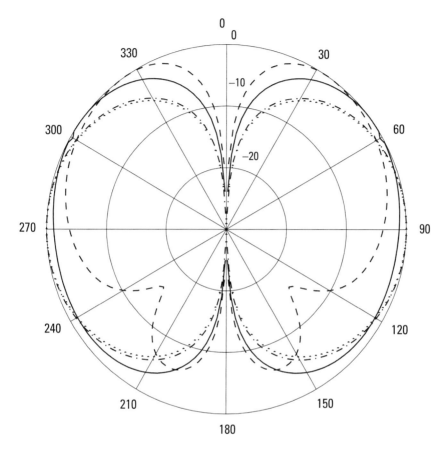

Figure 2.17 E plane radiation pattern for a vertical quarter-wave monopole mounted on a circular ground plane: '....' $t = 0$, '_ . _' $t = 0.2\lambda_0$, '_____' $t = 0.5\lambda_0$, '_ _ _' $t = 1\lambda_0$. The data was taken from [15] and normalized to a maximum gain of 0 dB.

If the ground plane is not infinitely thin, then charge can be stored along the circumference and the current may not be zero at the edges of the ground plane. This can alter the impedance and the radiation pattern of the antenna. Thus a monopole mounted on a conducting box will have different properties compared with a monopole mounted on a thin, finite metal sheet. This has implications for the handsets of cellular telephone antennas. The presence of the user's hand on the chassis of a cellular telephone handset can change the impedance of the antenna, the radiation pattern, and the radiation efficiency.

2.7 Simple Switched-Parasitic Antennas

Another method of providing an effect similar to the image of a dipole antenna, as shown in Figure 2.12, is to use a near-resonant half-wavelength parasitic wire element near an active element. This effect is similar to an element near a ground plane, in that the current in the parasitic wire is such that radiation away from the parasitic element is maximized. If the distance between the parasitic element and the driven element is small, then the current in the driven element I_2 is almost equal to I_1, but with 180° phase shift. That is, $I_1 \cong -I_2$. The radiation from the two elements is therefore canceled almost completely. The radiation efficiency is then very poor. The same is true when a dipole is placed very close to and parallel to a ground plane. The use of parasitic elements to reflect radiation from a driven element is important and plays a fundamental role in both switched parasitic antennas and Yagi-Uda antenna arrays, where the length of the element is critical to the directional characteristics of the array.

In Section 2.4, the impedance matrix method of calculating the currents in the parasitic elements of an array was given by (2.38) and (2.40). These equations showed that the mutual impedance and the self-impedance of each element in the array contribute to the current in every element.

2.7.1 Switched Active Antennas

Two different arrangements of dual-wire antennas, in which there is one driven dipole and one parasitic dipole, are shown in Figure 2.18(a) and 2.18(b), respectively [16]. These configurations are the simplest dipole parasitic arrays. The active element is driven with voltage V, and the parasitic element is marked S. Only the side-by-side configuration has a practical monopole implementation, and this is illustrated in Figure 2.18(c). The directional characteristics of the array can be controlled by activating the switch at S. When the switch is closed, the element resonates at the same frequency as the driven element if the two elements are of equal length. When the switch is open circuit, the element will not resonate at that frequency. This is the basis of an electrically controlled smart antenna. A dc electronic signal can control the switch setting and so controls the antenna directional characteristics. If it is then possible to switch the position of the active element between the two elements, this array constitutes a two-element wire switched active switched parasitic antenna (SASPA) [17].

The feed position located at the center of a dipole element and at the ground-plane end of the monopole element ensures that the change in

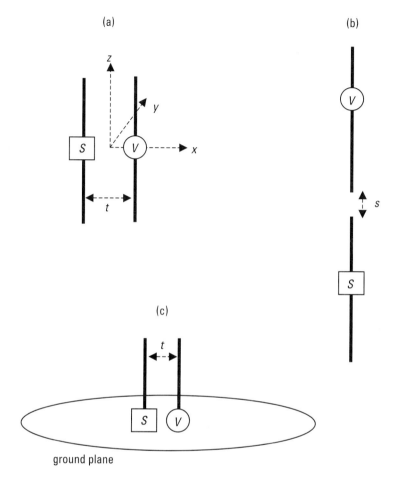

Figure 2.18 Two-wire switched parasitic antennas: (a) side-by-side dipoles, (b) collinear dipoles, and (c) side-by-side monopoles. *V* represents the driven element; *S* represents the switch in the parasitic element.

induced current is most significant. This position is not a necessity, as some switched parasitic antennas benefit from having the switch at a different location. A small change in the feed position away from the center of a dipole can change the input impedance of the antenna and can be used to optimize the impedance match to the transmission line, as shown by (2.33).

For simplicity, assume that the dipoles are oriented parallel to the z axis and lie in the xz plane. Both dipoles are $\lambda_0/2$ long, have a very small radius (i.e., less than $\lambda_0/100$), and are arranged in a side-by-side configuration

[Figure 2.18(a)]. When the switch is closed, the mutual impedance between the two elements is given in Figure 2.8. This provides sufficient information to calculate the four complex impedance elements in (2.40). If the input voltage $V = 1$ volt and S is closed, the currents in the driven and parasitic elements, I_V and I_S, respectively, can be calculated. The far-field radiation pattern is the sum of the radiated field from the two elements, providing the phase relationship between the two fields is taken into account. As the radiation pattern for a half-wave dipole is given by (2.28), the total electric field E_{tot} determined in the far field is:

$$E_{tot} = \frac{j\omega\mu}{2\pi rk} \frac{\cos[(\pi/2)\cos\theta]}{\sin\theta} \left[I_V + I_S e^{-jkt\cos\phi} \right] \qquad (2.41)$$

where t is the separation distance between the two dipole elements of the array. I_V and I_S are complex numbers representing the amplitude and phase of the currents in the driven and parasitic elements, respectively. The H plane radiation pattern lies in the xy plane and gives the ϕ dependence. This is significantly different from the circular pattern of an isolated dipole.

The H plane radiation pattern for a number of values of separation distance t are shown in Figure 2.19 for this two-dipole array. In all cases, the element on the right is the driven element, and the left element is the shorted parasitic element. The radiation patterns in Figure 2.19 have all been normalized for a maximum gain of 0 dB. The maximum radiated power lies in the direction of the driven element, that is, $\phi_{max} = 0°$ for $t = 0.1\lambda_0$ and $t = 0.25\lambda_0$. The parasitic element is therefore acting as a reflector in the array. The total radiated power is influenced by changes in the input impedance of the array relative to the characteristic impedance of the transmission line feed. This was discussed in Sections 2.1 and 2.3.

If the left element is driven and the right element is short-circuited, then $\phi_{max} = 180°$ and the radiation pattern given in Figure 2.19 is rotated through 180°. The antenna is a SASPA and the direction of the radiation is reversed by changing the functions of the two elements. This reversal can be achieved electronically using an RF switch to change the source from one antenna to the other, and by short-circuiting the other element to form a reflecting parasitic element. The drive circuit to control the radiation is given in Figure 2.20. Each dipole element has a switch fitted to the center of the wire (i.e., at the quarter-wavelength position for a half-wave dipole). When the RF source is applied to one element, this center switch is open, and the center switch in the other element is closed. With a single control line

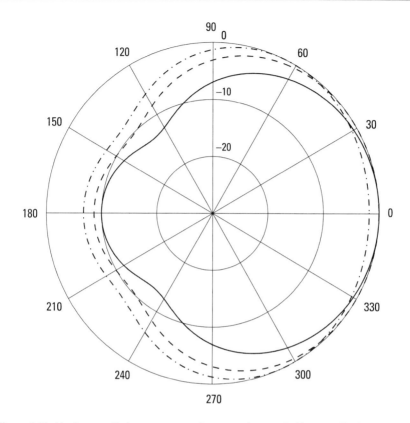

Figure 2.19 H plane radiation patterns of a two-element half-wave dipole switched active switched parasitic antenna (SASPA). The gain is expressed in decibels and the element spacing t varies from $0.1\lambda_0$ to $0.4\lambda_0$: '____' $t = 0.1\lambda_0$, '_ _ _' $t = 0.25\lambda_0$, '_ . _' $t = 0.4\lambda_0$.

driving the RF switch and two dc switches located at the center of each element, the direction of maximum radiation intensity can be reversed electronically. From symmetry the input impedance of the antenna is identical for the two switch positions.

2.7.2 Switched Parasitic Antennas

A fixed active switched parasitic antenna (FASPA) is an alternative switched parasitic antenna in which the beam direction can be controlled in a way similar to the SASPA. In this case, the RF source remains connected to a fixed active element and the state of the surrounding parasitic elements is

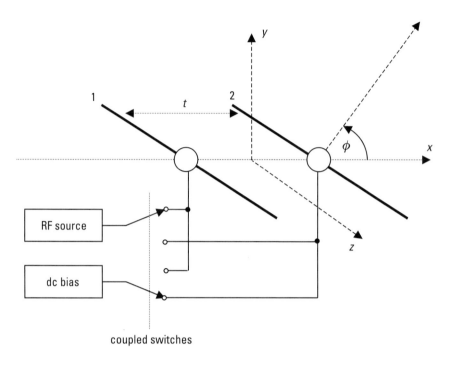

Figure 2.20 Control circuit to change the radiation direction of a two-element switched active antenna (SASPA). When the RF power is connected to element 1, the center element switch is open circuit and the center switch for element 2 is closed.

changed using dc switches. The simplest implementation of this technique is a three-element array, shown schematically in Figure 2.21. In this case, the central element, element 2, is always active, and when the switch in element 1 is closed and the switch in element 3 is open, $\phi_{max} = 0°$. When the switching is reversed, the direction of radiation is reversed and $\phi_{max} = 180°$. As with the SASPA, the input impedance of the FASPA is independent of the two switch positions, providing symmetry is retained. If it is assumed that the open-circuit element carries no current, then the antenna characteristics are identical to a two-element SASPA.

When the center switch of a half-wavelength parasitic element is closed, the element is near resonance and large currents can be induced. In this situation, the radiation and impedance characteristics of the array can be influenced significantly by the parasitic element. When the switch is open, the element is not close to resonance and the induced current is relatively small. In this case, the element does not play a significant role in the

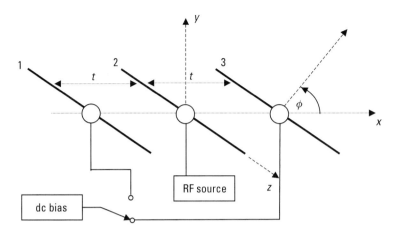

Figure 2.21 Schematic diagram of the configuration and circuitry of a three-element fixed active switched parasitic antenna (FASPA).

performance of the array. The difference in the H plane radiation pattern obtained using NEC [8] between these two cases is shown in Figure 2.22. As before, the antenna elements are dipoles of length $l = 0.05$m and radius $a = 0.001$m. The element spacing $t = 0.02$m. The gap in the open-circuit element is 0.001m and the applied frequency is 3 GHz. The result of the open-circuit parasitic element is a small increase in the gain of the three-element FASPA (0.5 dB) and a front-to-back ratio of 6.3 dB, compared with 5.2 dB for the two-element SASPA.

The effect of the open-circuit parasitic antenna element on the impedance of the feed element is quite small. Figure 2.23 shows the S_{11} when the spacing between the two antennas is $0.2\lambda_0$. There is very little shift in the resonant frequency and the 10-dB impedance bandwidth, for the SASPA is approximately 80 MHz, compared with 90 MHz for the FASPA.

The addition of a switched parasitic element in the collinear configuration [Figure 2.18(b)] for a half-wave dipole array has only a minor effect on both the E plane and the H plane radiation patterns when compared with a single driven dipole. This is because the current in the parasitic element is quite small for spacings $t > 0.4\lambda_0$. When $t < 0.4\lambda_0$, the phase difference between the induced current in the parasitic element and the current in the driven element is so small that very little asymmetry is evident in the E plane radiation pattern. The net effect is to increase the beamwidth of the broadside radiation pattern. For these reasons, the collinear configuration serves no useful purpose in switched parasitic antenna arrays.

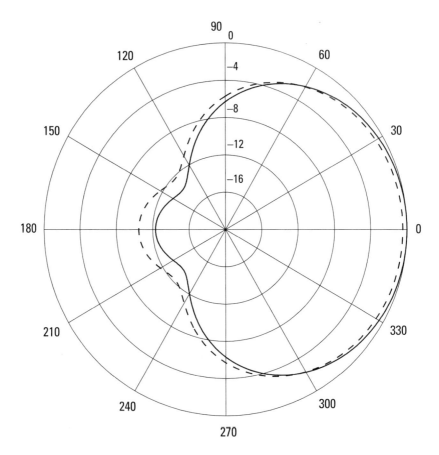

Figure 2.22 Comparison between the H plane radiation pattern of a two-element SASPA and a three-element FASPA. The gain is plotted in dB, and has been normalized to the maximum gain of the three-element array. '____' SASPA, '_ _ _' FASPA.

The two examples of switched parasitic antennas given in Figures 2.20 and 2.21 are linear arrays, that is, the active and parasitic monopoles or dipoles lie in a line. SASPA and FASPA antennas can be constructed by employing radial symmetry. For example, a driven monopole circled by switchable parasitic monopoles can be used to scan the beam through 360° [17, 18]. Providing symmetry is retained in the antenna design and the switch configuration, the input impedance of the antenna remains independent of the switch position and so is independent of the main beam direction of the antenna. This is an important consideration in the operation of

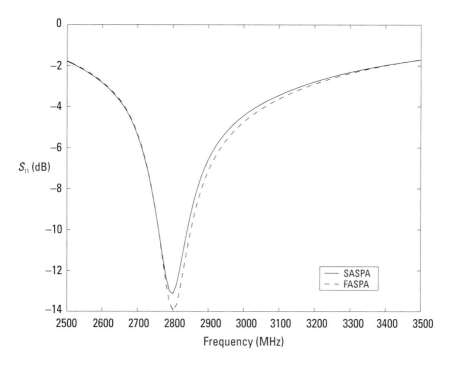

Figure 2.23 S_{11} (dB) variation plotted as a function of frequency for the two-element switched active antenna (SASPA) and the three-element switched parasitic antenna (FASPA).

scanning arrays. More complex switched parasitic wire arrays are discussed in Chapter 4.

2.8 Wire Element Phased Arrays

Phased arrays of wire elements usually consist of a uniform array of resonant elements. In one dimension, this can be a straight line or circle of elements, and in two dimensions, a rectangular grid. All antenna elements in the array are driven, and the phase and magnitude of the RF voltage applied to each element is controlled electronically. This allows the direction of the main beam to be controlled electronically.

A simple linear phased array consists of dipoles arranged either side by side or in the collinear configuration (see Figure 2.7). For the side-by-side case, the far-field radiation pattern for n elements is given by [1, 2]:

$$E_{tot} = \frac{j\omega\mu}{2\pi rk} \frac{\cos[(\pi/2)\cos\theta]}{\sin\theta} \sum_{i=1}^{n} I_i e^{-jkt_i \cos\phi} \qquad (2.42)$$

where I_1, I_2, ..., I_n are the magnitudes of the currents in the elements and t_1, t_2, ..., t_n, are the separation distances between the first element and all other elements. The θ dependence terms are identical to the θ angular dependence of the elements alone, given by (2.28). This is called the element factor. Each term in the summation in (2.42) describes the influence of one element in the array, and the total effect is called the array factor (AF) [1, 2].

For example, a cellular telephone base-station antenna may be constructed from three collinear elements forming a phased array. The voltage applied to each element has the same magnitude, but there is a sequential phase shift ξ between adjacent elements determined from the element spacing and the required angle θ_{max} below the horizontal plane.

Consider a three-element side-by-side phased array in which the elements have the same length. If the elements are equally spaced with a separation distance t, then:

$$t_1 = 0, \; t_2 = t, \text{ and } t_3 = 2t \qquad (2.43)$$

The simplest form of analysis is to assume that the magnitude of the voltage V applied to each element is identical, and that there is no mutual coupling between the elements. The currents can be written as:

$$I_1 = V/Z_{11} \qquad I_2 = Ve^{-j\xi}/Z_{11} \qquad I_3 = Ve^{-2j\xi}/Z_{11} \qquad (2.44)$$

where ξ is the phase difference between the voltages applied to adjacent elements.

If the antenna elements are close together, the assumption that the current in one element does not influence the current in the other may be false, and mutual coupling must be considered. For the side-by-side configuration, the data in Figure 2.8 suggests that this distance is less than $2\lambda_0$. Figure 2.24 shows the H plane radiation patterns for the three-element array, calculated with and without mutual coupling, for the two different separations $t = 0.2\lambda_0$ and $t = 0.4\lambda_0$. In Figure 2.24(a) the constant phase difference between adjacent elements is $\xi = 0°$, and in Figure 2.24(b), $\xi = 45°$. For $\xi = 0°$, the direction of maximum radiation is perpendicular to the array at $\phi_{max} = 90°$, commonly referred to as the broadside direction, and for $\xi = 45°$, the radiation is mainly parallel to the array (i.e., $\phi_{max} = 0°$). This is referred to as the end-fire direction.

Table 2.2 gives the values of the currents I_{1c}, I_{2c}, I_{3c}, I_{1u}, I_{2u}, and I_{3u}, representing the coupled and uncoupled currents of the three elements calculated, assuming the voltage applied to each element has a magnitude of 1 volt. The current values are in mA and the phase is in degrees.

Table 2.2 shows that the uncoupled current values in each element are identical, but for the coupled values, the center current is significantly different. For this reason, the normalized radiation patterns given in Figure 2.24 for uncoupled and coupled calculations are significantly different. The maximum effect is evident at $\phi = 0°$, for $\xi = 45°$, the reverse direction for the array. The front-to-back ratio is significantly different in the case of the phase-shifted feed. This effect is exacerbated when the characteristic impedance of the feed lines is taken into account, as a poor value of S_{11} can significantly alter the magnitude of the current in each element. It is also imperative that the phase shifters do not significantly alter the magnitude of the signal applied to the element, and that the attenuators/amplifiers for each element do not induce a phase shift. While the input impedance of each element in the array remains fixed for all switch positions, the matching of each element to the transmission line feed depends very much on its location in the array and the proximity of the adjacent elements. The design and fine-tuning procedure for a phased array is more complex than for an equivalent switched parasitic antenna.

For an n element, uniform linear array of side-by-side elements having identical length, the normalized array factor AF is given by [1, 2]:

Table 2.2

Currents in a Three-Element Phased Array, with Element Separations Given by t^*

$t (\lambda_0)$	ξ (°)	I_{1c} (mA)	I_{2c} (mA)	I_{3c} (mA)	I_{1u} (mA)	I_{2u} (mA)	I_{3u} (mA)
0.2	0	$8.3 - 1.7j$	$9.2 + 3.8j$	$8.3 - 1.7j$	$10.2 - 6.0j$	$10.2 - 6.0j$	$10.2 - 6.0j$
0.4	0	$12.2 - 1.9j$	$15.7 - 3.7j$	$12.2 - 1.9j$	$10.2 - 6.0j$	$10.2 - 6.0j$	$10.2 - 6.0j$
0.2	45	$8.3 - 1.7j$	$9.2 - 3.8j$	$4.7 - 7.1j$	$10.2 - 6.0j$	$3.0 - 11.5j$	$-6.0 - 10.2j$
0.4	45	$12.2 - 1.9j$	$8.4 - 13.7j$	$-1.9 - 12.2j$	$12.2 - 1.9j$	$7.3 - 10.0j$	$-1.9 - 12.2j$

* Currents with the subscript c include the effects of mutual coupling, and currents with the subscript u assume no mutual coupling effects.

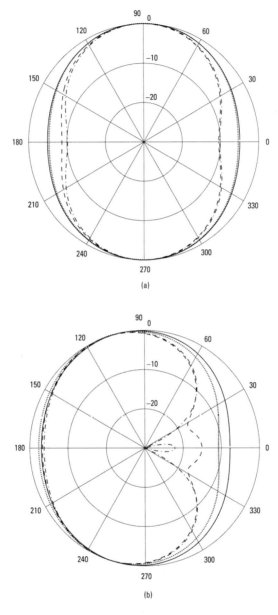

(a)

(b)

Figure 2.24 H plane radiation patterns (dB) with and without mutual coupling for a three-dipole side-by-side phased array with separation distances $t = 0.2\lambda_0$ and $0.4\lambda_0$: (a) the phase shift $\xi = 0°$, (b) $\xi = 45°$. '____' $t = 0.2\lambda_0$ with mutual coupling, '....' $t = 0.2\lambda_0$ without mutual coupling, '_ _ _' $t = 0.4\lambda_0$ with mutual coupling, and '_ . _' $t = 0.4\lambda_0$ without mutual coupling.

$$AF = \frac{\sin(n\psi/2)}{n\sin(\psi/2)} \qquad (2.45)$$

where the total phase shift between adjacent elements is $\psi = kt\cos\phi + \xi$. This term includes both the physical separation between the elements and the applied voltage phase shift ξ. This is the sinc function found in (2.24), which is the Fourier transform of a single square wave pulse. In deriving (2.45), it is assumed that the effect of mutual coupling is insignificant and so the array of identical wire elements forms a spatial square wave of uniform radiators.

If the spacing t of the array elements is fixed, and the phase shift between the elements ξ is varied continuously over the range from $0°$ to $360°$, the main-beam direction is rotated through the ϕ plane. The main disadvantage of this type of array is that the radiation pattern is symmetrical about the line of the array, and so if ϕ_{max} is the direction of the main beam, there is an identical beam at $-\phi_{max}$ and the effective front-to-back ratio is not good. There are also some values of ξ in which the sidelobe levels can be as high as the main-beam radiation level. Various optimization strategies used to improve the performance of linear arrays include the use of nonuniform spacing of the elements and nonuniform voltage amplitudes applied to individual elements [1, 2].

A circular phased array consists of a set of equally spaced elements on the perimeter of a circle [2, 19]. It can be used to sweep the ϕ_{max} through $360°$ while maintaining the same front-to-back ratio and sidelobe levels. The phase and amplitude of each element must be controlled independently to reduce the sidelobe levels [2, 19].

An alternative approach is to use a two-dimensional array of elements, arranged in a rectangular grid. This is referred to as a planar array; the direction of the main beam can be varied in both the θ and ϕ directions simultaneously [2].

Thus, linear, circular, and planar arrays can provide flexibility in beam scanning; however, the cost is additional hardware such as electronically controlled phase shifters, attenuators, and amplifiers. Significant computer support is required to control the phase and amplitude of the radiation from each element.

In dual-frequency operations, the phase-shifter settings required to direct the beam will not be the same for both frequencies, as the kt factor in ψ in (2.45) is wavelength dependent. This creates significant problems for standard phased-array designs operating at more than one frequency. The problem can be overcome through the use of time-delay elements rather than phase-delay elements. The frequency dependence and impedance of the time-delay elements must be carefully matched to ensure optimal results. This process can be quite time consuming if large numbers of radiating elements are involved. The recent trend is toward digital radio frequency systems in which the RF signal carrier is generated independently at each element in the array, and the required phase shifts, time delays, and amplitude variations are controlled by digital logic located at the array elements individually. These arrays are referred to as active arrays, and have been described as the next generation of phased antenna arrays. The problems associated with mutual coupling between the elements still remain, and must be taken into account for precision beam positioning.

2.9   Dual-Band Wire Antennas

In this chapter, the wire antennas discussed have all been considered to have a single operating resonant frequency f_0. This frequency is determined by the ability of the antenna to match a 50Ω transmission line, and corresponds to the frequency when the S_{11} value is a minimum. In the case of a dipole in free space, the length of the antenna is approximately $\lambda_0/2$. As the frequency increases, there are other frequencies when this antenna is resonant; that is, when the S_{11} is a minimum. This occurs at frequencies close to the odd harmonics of this base frequency (i.e., at $3f_0$, $5f_0$, $7f_0$, $9f_0$, and so on). The reflection coefficient of a thin wire dipole over an extended frequency range is shown in Figure 2.25.

There are at least two reasons why it is not desirable to operate an antenna at two of these harmonic frequencies simultaneously. First, at the higher frequencies, the E plane radiation pattern has a number of deep nulls, and so the main-beam direction θ_{max} and beamwidth B_θ at the two frequencies is significantly different. Figure 2.26 illustrates the radiation patterns for the first (2.85 GHz) and second (8.80 GHz) resonant frequencies of the dipole described in Figure 2.25. The second reason is that the input stages of the receiver and the spectral purity of a high-powered transmitter are usually sensitive to these harmonics. This means that the possibility of cross-talk between the two channels is high. For these two reasons, dual-channel systems are not usually designed for the harmonic frequencies.

There are other techniques that achieve a dual resonance at frequencies f_1 and f_2, in a single-wire, monopole antenna structure. The effective electrical length of the wire structure must be different at different frequencies. In

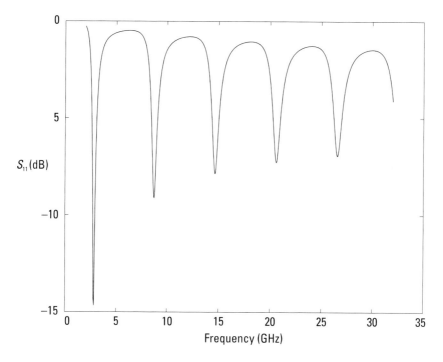

Figure 2.25 S_{11} (dB) for a center-fed dipole of length 0.05m, radius 0.0001m, over a wide frequency range.

Section 2.7, the effective length of a parasitic dipole antenna was changed by an electronic switch at its center. The change was from a resonant length $\lambda_0/2$ to a nonresonant length $\lambda_0/4$ at the frequency of interest f_0, and so the contribution of the new wire structure to the overall radiation characteristics is insignificant. The open-circuited wire element has length $\lambda_0/4$ and resonates at the even harmonic of the base frequency with the lowest frequency of $2f_0$.

In Figure 2.27, a lumped-impedance element Z_x has been inserted at a distance d from the feed point at the base of the monopole. This is referred to as a loaded wire antenna [20] and can be used to allow the element to resonate at two different frequencies that are relatively closely spaced. If these two frequencies are within one bandwidth of each other, then the impedance bandwidth of the element about a single center frequency has effectively doubled.

Consider the performance of a loaded monopole antenna at two frequency bands with center frequencies f_1 and f_2. If $f_1 < f_2$, then the effective length of the antenna for f_1 is longer than that required for the resonant fre-

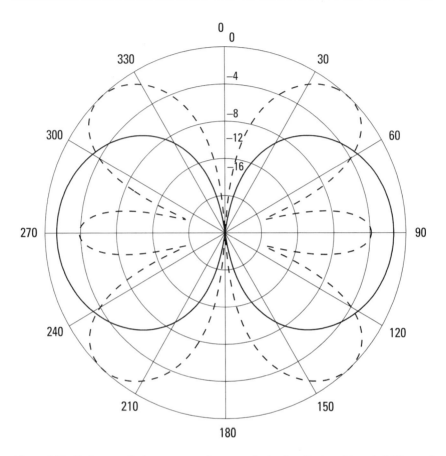

Figure 2.26 E plane radiation patterns for a vertical wire dipole of length 0.05m and radius 0.0001m at 2.85 GHz and 8.80 GHz. The gain for both plots has been normalized to 0 dB for the 8.8-GHz pattern. '_____' 2.85 GHz, '_ _ _' 8.8 GHz.

quency f_2. The lumped-impedance method requires first that the length l_1 of the total antenna must approximately correspond to $\lambda_1/4$, and second, that the lumped-impedance Z_x is approximately zero at this frequency. At the second frequency, the lumped impedance should be very large, and the distance to the element from the feed point should be approximately $\lambda_2/4$. It is also desirable that the antenna is not resonant at frequencies outside these two frequency bands.

If the overall length of the monopole is 0.25m and no load resistor is present, then $f_1 \cong 300$ MHz. With a load placed at 0.183m from the base,

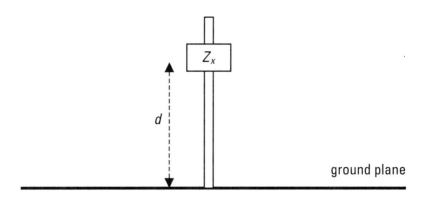

Figure 2.27 Design technique to achieve a dual-band wire monopole antenna on a ground plane using a passive lumped-impedance element Z_x in the wire at position d.

and if this load is successful in reflecting most current at f_2, then the resonant frequency is increased to $f_2 \cong 410$ MHz. The transition between f_1 and f_2 occurs when the load resistance changes from 100Ω to 800Ω. The variation of the resonant frequency is shown as a function of load impedance in Figure 2.28 for a purely resistive load. This data was calculated using the NEC code [8], and the resonant frequency was selected to be the frequency with a minimum VSWR using a 50Ω feed line. The calculations were repeated at higher frequencies with appropriately scaled-down lengths, and the impedance transition occurred between 100Ω and 800Ω.

For a dual-frequency antenna resonant at f_1 and f_2, the lumped impedance should meet the criteria:

$$|Z_x(f_1)| < 100\Omega \text{ and } |Z_x(f_2)| > 800\Omega \qquad (2.46)$$

The simplest lumped element that shows this frequency trend is a purely inductive element. The impedance is given by:

$$Z_x(f) = jL2\pi f \propto f \qquad (2.47)$$

In this case, the rate of change of Z_x with f is relatively small, and so this solution is not adequate unless the difference between f_1 and f_2 is large. From (2.47) and Figure 2.28, the condition $f_2 = 8f_1$ is required, and this is unsuitable for most communications systems. The desirable frequency bands are usually much closer together.

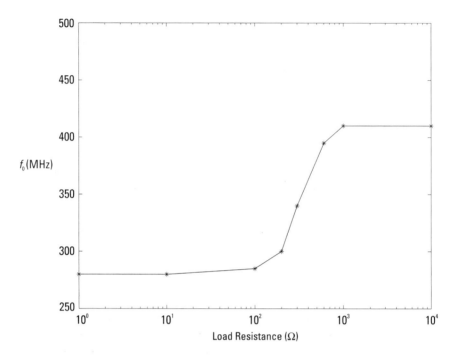

Figure 2.28 Variation in the resonant frequency of a monopole antenna (length 0.25m, radius 5 mm) as a function of the series load resistance located 0.183m from the base.

In order to increase the variation of the load impedance Z_x with frequency, it is necessary to design a two-port filter with a larger number of poles and/or zeros. The design methodology for such filters can be found in circuit analysis books [21]. Workable designs usually involve a tuned circuit arranged for band pass characteristics. It is also important to note that, for this application, not only the magnitude of the change from $Z_x(f_1)$ to $Z_x(f_2)$ is important, but also the phase change introduced into the antenna element by the filter.

Figure 2.29 gives an example of the final design of a filter for a 900/1,200 MHz monopole antenna. The total length of the monopole is 0.07m, with a wire radius of 0.001m. The lumped impedance placed at $d =$ 0.048m consists of a parallel arrangement of a resistor (1 kΩ), a capacitor (4 pF), and an inductor (5 nH). The circuit is given in Figure 2.29(a), and the impedance characteristics of the circuit are given in Figure 2.29(b). The magnitude of the impedance change from 900 MHz to 1,200 MHz does not

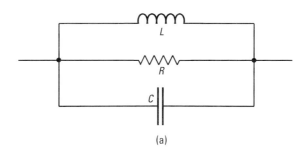

(a)

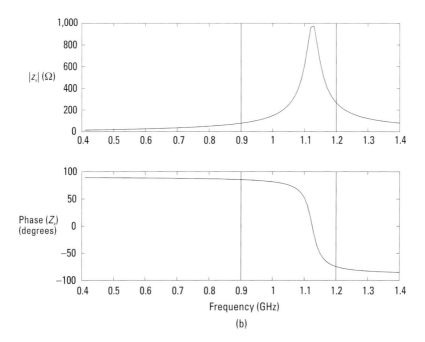

(b)

Figure 2.29 (a) Lumped-impedance circuit diagram for a dual-band monopole operating with center frequencies 900 MHz and 1,200 MHz. (L = 5 nH, R = 1 kΩ, and C = 4 pF). (b) Lumped-impedance magnitude and phase plotted as a function of frequency.

satisfy (2.45), but there is a significant phase transition between the two impedance values to ensure the two resonant frequencies are well defined. Figure 2.30(a) shows the S_{11} characteristics versus frequency, and Figure 2.30(b)

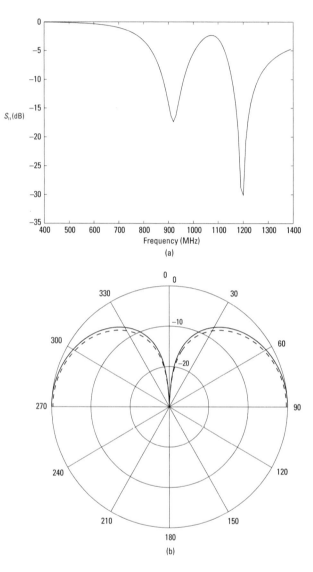

Figure 2.30 Dual-band monopole on an infinite ground plane: (a) S_{11} (dB) across the frequency range; and (b) E plane radiation patterns (dB) at 900 MHz and 1,200 MHz. '____' 900 MHz, '_ _ _' 1,200 MHz.

gives the E plane radiation pattern for the antenna at the two center frequencies. The impedance bandwidth is approximately 9% at both frequencies. The two radiation patterns are very similar, implying that the antenna coverage at the two frequencies is virtually identical.

The design of a dual-band, lumped-impedance monopole antenna requires the optimization of its overall length l, wire radius a, lumped-impedance position d, and the R, L, and C components of the impedance itself. The R component in the design influences the impedance bandwidth at both frequencies. An increase in C produces a significant decrease in the upper resonant frequency f_2. An increase in L decreases both f_1 and f_2; the overall length of the monopole affects the lower resonant frequency f_1. The position of the impedance element d affects the upper frequency value. An increase in the wire radius a increases the impedance bandwidth but also shifts the resonant frequencies f_1 and f_2. It is also important to note that the design must be arranged to ensure that thermal effects on the circuit components will not have a significant effect on the filter characteristics at the two frequencies. Thus, the design of a practical antenna requires an assessment of the sensitivity of the antenna performance to small changes in each design parameter. This is a complex multiparameter optimization problem, and various approaches to solving such problems are discussed in Chapter 5.

2.10 Dual-Band Switched Parasitic Wire Antennas

The elements required for a dual-band switched parasitic antenna must be designed with two goals. The elements must be resonant at both frequencies when the control switch is closed, and they must be nonresonant at both frequencies when the switch is open. In addition, it is desirable that the radiation pattern of the array at both frequencies does not change significantly. This means that the area of coverage of the antenna system is approximately the same.

Following from the discussion on dual-band wire antennas (Section 2.9), the most likely candidate as an element in a switched parasitic dual-band antenna is the loaded monopole or dipole located in the vicinity of some switched parasitic elements. Figure 2.31(a) is a schematic diagram of a dual-band FASPA. The center element is always the driven element. The values of $L = 5$ nH and $C = 4$ pF are identical to that used for the parallel loaded monopole designed for dual-band operation in Figure 2.29; however, the value of R was increased to 10 kΩ to achieve an improved bandwidth [see Figure 2.31(b)]. The element spacing is 0.045m, and both f_1 and f_2 are close to 900 MHz and 1,200 MHz, respectively. The length of the center element was unchanged at 0.7m, but the parasitic monopoles were increased to 0.08m to achieve an improved front-to-back ratio. The difference in length between the elements is related to the Yagi-Uda antenna design discussed in

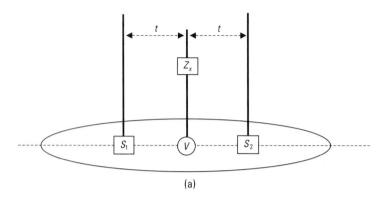

(a)

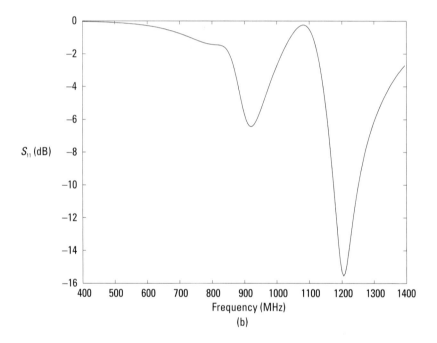

(b)

Figure 2.31 (a) Schematic diagram of the dual-band, three-element, switched parasitic antenna. (b) S_{11} (dB) variation as a function of frequency for the antenna in (a).

Section 2.11. The wire radius for all three elements $a = 0.001$m. The switches S_1 and S_2 in Figure 2.31(a) are alternatively closed or open to change the direction of the radiation through 180°.

The E plane and H plane radiation patterns for this antenna are plotted in Figure 2.32 for both resonant frequencies. The gain has been plotted in decibels, normalized to a maximum gain of 0 dB at 900 MHz. The front-to-back ratio for the antenna at both frequencies is greater than 9 dB. The design assumes a ground plane of infinite extent, and all results presented in Figures 2.31 and 2.32 were calculated using NEC [8].

2.11 Yagi-Uda Wire Antennas

An antenna array of dipoles in the side-by-side configuration, in which there is one driven element and the parasitic elements are shorter and longer than the driven element, is commonly referred to as a Yagi-Uda array [1, 2], having been named after its inventors [22, 23]. The driven element is approximately one-half-wavelength long. The surrounding parasitic elements on one side are slightly shorter than the driven element and are called directors. The elements on the other side are slightly longer than the driven element and are called reflectors. The maximum radiation is in the direction of the directors, and a narrow beamwidth is achieved through the use of a large number of directors [1, 2]. The directors also contribute to the antenna performance by significantly increasing the impedance bandwidth. A typical Yagi-Uda antenna is shown schematically in Figure 2.33. One common application of these antennas is for television reception.

Highly directional Yagi-Uda antennas can be designed with a large bandwidth. The design and optimization procedures using the MOM and other techniques are commonly available [1, 2].

A modification of the Yagi-Uda concept using switched parasitic elements was patented by Gueguen in 1974 [24]. The feed element was a monopole on a ground plane. This was surrounded with a number of circles of monopoles arranged as a radial array of Yagi-Uda antennas (i.e., the length of the elements decreased away from the active element). The array could be directed electronically by changing the impedance state at the feed point of these monopole elements.

An alternative approach to using the Yagi-Uda concept in switched parasitic antennas is to change the effective length of the parasitic elements by switching in reactive elements [25–28]. In this way it is possible to change a director element to a reflector element in the array, and so change the direction of radiation. This has the effect of increasing the front-to-back ratio for the antenna, in comparison to an antenna in which all elements are of equal length.

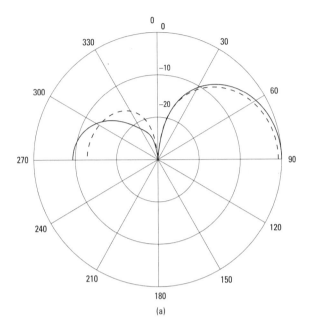

(a)

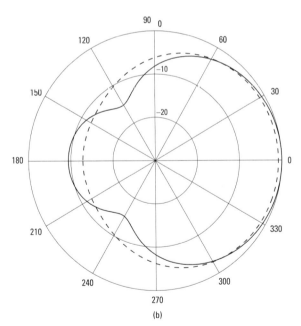

(b)

Figure 2.32 (a) E plane radiation pattern (dB) and (b) H plane radiation pattern (dB), for the dual-band, switched parasitic antenna using three monopoles. '___' 900 MHz, '_ _ _' 1,200 MHz.

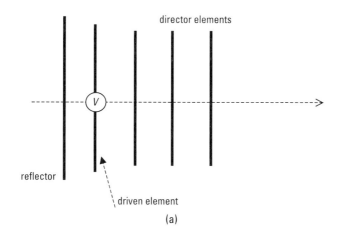

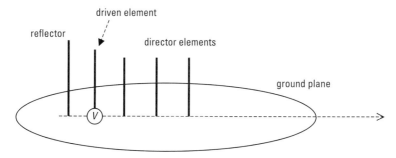

Figure 2.33 Typical Yagi-Uda antenna configuration of (a) dipole elements and (b) monopole elements on a ground plane.

The Yagi-Uda concepts have also been applied to combinations of slot and wire antennas [11], to slot antennas [29], and, more recently, to patch antennas [30]. This last implementation is discussed further in a switched parasitic configuration in Chapter 3.

2.12 Summary

In this chapter we discussed the fundamental principles of wire antennas, using three important descriptors of antenna performance—input impedance as measured by the magnitude of the reflection coefficient S_{11}, the beamwidth, and the bandwidth.

The basic wire-radiating elements commonly used in antennas are half-wave dipoles in free space and quarter-wave monopoles mounted on a ground plane. A combination of these elements can be used to improve the directional characteristics and the impedance match of the antenna array. Wire elements are used in phased arrays, switched parasitic antennas, and Yagi-Uda antennas. Both phased arrays and switched parasitic antennas can be electronically controlled to change the direction of the radiation. Two-dimensional phased arrays allow two-dimensional beam scanning, and switched parasitic antennas usually offer only one-dimensional scanning. The advantages of switched parasitic antennas include the relatively small size, constant input impedance, multiband operation, and simple directional control using electronics. The performance of the FASPA is usually slightly better than that of the SASPA, as the open-circuit parasitic element is slightly directive. The FASPA does not require the inclusion of an RF switch, which can be expensive for high-powered transmitting applications. The size of the footprint of the SASPA is less than that required for the equivalent FASPA.

In Chapter 3, switched parasitic patch antennas are discussed. Chapter 4 contains a number of more complex switched parasitic antenna designs, both wire and patch. In the later sections of Chapter 2, it became evident that the number of variables required in the design of a multiband, multielement, electronically steerable antenna system is quite large. The optimal design of such antennas requires multiparameter optimization techniques, and these are discussed in Chapter 5.

In this chapter, the switch required to change a resonant wire element to a nonresonant wire element has been assumed to have ideal performance; that is, zero resistance in the closed position and zero capacitance in the open position. This is sometimes not the case, and in Chapter 6, the degradation in performance of switched parasitic antennas resulting from nonideal behavior is considered.

References

[1]   Stutzman, W. L., and G. A. Thiele, *Antenna Theory and Design*, New York: John Wiley and Sons, 1981.

[2]   Balanis, C. A., *Antenna Theory Analysis and Design*, 2nd ed., New York: John Wiley and Sons, 1997.

[3]   Pozar, D. M., *Microwave Engineering*, 2nd ed., New York: John Wiley and Sons, 1998.

[4] Smith, G. S., *An Introduction to Classical Electromagnetic Radiation,* Cambridge, U.K.: Cambridge University Press, 1997.

[5] Moller, K. D., *Optics,* Mill Valley, CA: University Science Books, 1988.

[6] Ramo, S., J. R. Whinnery, and T. van Duzer, *Fields and Waves in Communications Electronics,* 2nd ed., New York: John Wiley and Sons, 1984.

[7] Neff, H. P., *Basic Electromagnetic Fields,* 2nd ed., New York: Harper and Row, 1987.

[8] NEC (Numerical Electromagnetics Code). This code was originally developed at the Lawrence Livermore National Laboratories, California, and now is available from R. Anderson at the Web site http://www.qsl.net/wb6tpu/swindex.html, 2000.

[9] Burberry, R. A., *VHF and UHF Antennas,* London: Peter Peregrinus, 1992.

[10] Schelkunoff, S. A., and H. J. Friis, *Antennas Theory and Practice,* New York: John Wiley and Sons, 1952.

[11] Thiel, D. V., "Optimised Slot Reradiation to Modify Foreground Reflection into an Array," *Proc. IEE,* Vol. 120, September 1973, pp. 962–964.

[12] Griffiths, J., *Radio Wave Propagation and Antennas: An Introduction,* Englewood Cliffs, NJ: Prentice Hall, 1987.

[13] Saario, S., et al., "Effect of Air Gaps in RF Current Suppression Using Ferrite Beads," URSI National Radio Science Meeting, University of Colorado, Boulder, Colorado, January 1998, p. 15.

[14] Richmond, J. H., *Monopole Antenna on Circular Disk,* Technical Report 711639-1, Electro Science Laboratory, Ohio State University, July 1979.

[15] Weiner, M. M., et al., *Monopole Elements on Circular Ground Planes,* Norwood, MA: Artech House, 1987.

[16] Thiel, D. V., S. G. O'Keefe, and J. W. Lu, "Electronic Beam Steering in Wire and Patch Antenna Systems Using Switched Parasitic Elements," *IEEE Antennas and Propagation Symposium,* Baltimore, MD, 1996, pp. 534–537.

[17] Preston, S., D. V. Thiel, and J. W. Lu, "A Multibeam Antenna Using Switched Parasitic and Switched Active Elements for Space-Division Multiple Access Applications," *IEICE Trans. Electronics,* Vol. E82-C, July 1999, pp. 1202–1210.

[18] Sibille, A., C. Roblin, and G. Poncelet, "Circular Switched Monopole Arrays for Beam Steering Wireless Applications," *Electronics Letters,* Vol. 33, March 1997, pp. 551–552.

[19] Davies, D. E. N., "Circular Arrays," in *Handbook of Antenna Design,* Vol. 2, A. W. Rudge, et al. (eds.), London: Peter Peregrinus, 1982.

[20] Boag, A., et al., "Design of Electrically Loaded Wire Antennas Using Genetic Algorithms," *IEEE Trans. Antennas and Propagation,* Vol. 44, May 1996, pp. 687–695.

[21] Ellis, M. G., *Electronic Filter Analysis and Synthesis,* Norwood, MA: Artech House, 1994.

[22] Uda, S., "Wireless Beam for Short Electric Waves," *J. IEE* (Japan), March 1926, pp. 273–278, and November 1927, pp. 1209–1219.

[23] Yagi, H., "Beam Transmission of Ultra-Short Waves," *Proc. IRE*, Vol. 26, June 1928, pp. 715–741.

[24] Gueguen, M., Electronically Step-by-Step Rotated Directive Radiation Beam Antenna, U.S. Patent No. 3846799, November 1974.

[25] Harrington, R. F., "Reactively Controlled Directive Arrays," *IEEE Trans. Antennas and Propagation*, Vol. 26, May 1978, pp. 390–395.

[26] Dumas, T. A., and L. V. Griffee, Electronically Rotated Antenna Apparatus, U.S. Patent No. 4631546, December 1986.

[27] Sibille, A., C. Roblin, and G. Poncelet, "Beam Steering Circular Monopole Arrays for Wireless Applications," *10th Intl. Conf. Antennas and Propagation, ICAP'97*, Edinburgh, U.K., Vol. 1, April 1997, pp. 358–361.

[28] Scott, N. L., M. O. Leanard-Taylor, and R. G. Vaughan, "Diversity Gain from a Single-Port Adaptive Antenna Using Switched Parasitic Elements Illustrated with a Wire and Monopole Prototype," *IEEE Trans. Antennas and Propagation*, Vol. 47, June 1999, pp. 1066–1070.

[29] Coe, R. J., and G. Held, "A Parasitic Slot Array," *IEEE Trans. Antennas and Propagation*, Vol. 12, January 1964, pp. 10–16.

[30] Huang, J., and A. C. Densmore, "Microstrip Yagi Array Antenna for Mobile Satellite Vehicle Application," *IEEE Trans. Antennas and Propagation*, Vol. 39, July 1991, pp. 1024–1030.

3

Patch Antennas

3.1 Radiation from a Patch Antenna

There are three common types of microstrip antennas: microstrip patch antennas, microstrip traveling-wave antennas, and microstrip slot antennas. Patch antennas are fabricated on one side of a dielectric sheet, with the other side being a continuous ground plane [1–5]. A number of advantages make patch antennas attractive. For example:

- They are simple to fabricate. Using photolithographic techniques, large numbers of elements can be made using the same number of processing steps as that required for one element. For this reason, patch antenna arrays are common.
- They can be integrated directly with electronics [6]. The dielectric sheet forms the printed circuit board (PCB) or the insulating layer in integrated circuits fabricated in semiconductor materials.
- The conformal surface of the antenna can be integrated unobtrusively into the surface of an aircraft or building and covered with a smooth protective coating.

A number of geometrical shapes are used for patch antennas (e.g., square, rectangle, circle, triangle, ellipse, rectangle with truncated corners, and pentagon). Parasitic patch antennas require significant coupling between adjacent patches; this can only be achieved when the edges of the two patches are close. In this book, we will direct attention to rectangular and square patches, as the sides are straight lines and adjacent elements can be arranged for maximum coupling. In addition, the transmission lines used to feed the active patches cannot be located on the same surface as the radiating elements. This common method of feeding patch antennas through microstrip transmission lines will not be considered in this book. For parasitic patch antennas, the preferred feed mechanism is via a coaxial probe (see Figure 3.1) or a microstrip line coupled to the driven element via a slot [see Figure 3.1(b)] [4]. In these two cases, the transmission lines lie on a plane different from the patch elements, and so do not interfere with the placement of parasitic elements.

A rectangular patch antenna resembles a short length of wide microstrip line open-circuited at each end. The dielectric region between the patch and the ground plane forms a parallel plate waveguide. The center conductor of the coaxial cable provides a current source that generates fields in this waveguide. These fields propagate radially out from the probe. At the edges of the patch there is a transition between the waveguide mode and propagation into the region beyond. This is a discontinuity, and so radiation occurs at this point.

If the length of the patch L is approximately $\lambda_g/2$, where λ_g is the wavelength in the dielectric-filled, parallel plate waveguide, the waveguide section is resonant and there is maximum power transfer to the antenna. This is the first-order mode if $L > W$, and has the lowest resonant frequency for the patch [1]. For thin dielectric slabs, $\lambda_g \cong \lambda_0/\varepsilon_r$. Because of fringing fields at the ends of the patch, the effective length of the structure L_{eff} is always greater that the actual length L. The mathematical relationship between L_{eff} and L is not a simple expression, and is dependent on both the relative permittivity ε_r, and the thickness t of the substrate. For this reason, L is generally found to be:

$$L \cong 0.49\lambda_g \tag{3.1}$$

At the ends of the patch, the fields are propagated in two ways: in a surface-wave mode trapped by the dielectric and as a radiated field. This radiation mechanism is equivalent to two parallel radiating slots located at the ends of the microstrip line. The width of each slot is equal to the patch width W. The center-to-center separation distance between the two slots is $L_{eff} = L + s$, where the equivalent slot thickness s is approximately equal to the

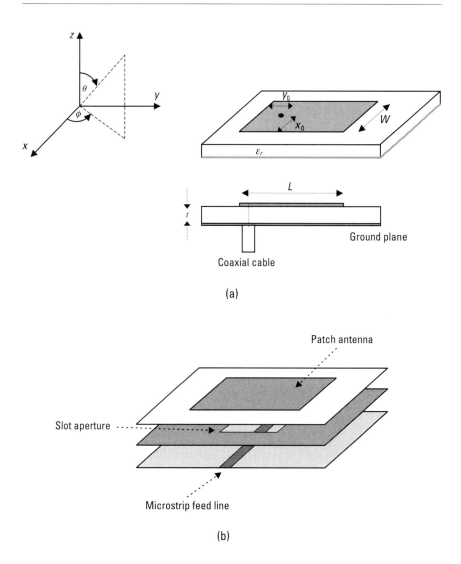

Figure 3.1 (a) Rectangular patch antenna with coaxial feed at (x_0, y_0); (b) slot-fed patch antenna showing the three conducting layers separated by two dielectric layers.

dielectric thickness t. The separation distance between the two slots is approximately $0.5\lambda_g$. This is the two-dimensional equivalent of the half-wave dipole described in Section 2.1 and is referred to as a half-wave patch antenna [2].

A rectangular patch antenna has a number of resonant modes in the y direction with frequencies f_m given by the equation [4]:

$$f_m = \frac{mc}{2L_{eff}\sqrt{\varepsilon_r}}$$

(3.2)

where m is an integer representing the mode number ($m = 1, 2, 3, \ldots$), c is the velocity of light in a vacuum, and ε_r is the relative permittivity of the dielectric slab.

There are also resonances f_n in the x direction, where:

$$f_n = \frac{nc}{2W\sqrt{\varepsilon_r}}$$

(3.3)

The radiation pattern for the dominant mode can be calculated from the two equivalent slots separated by the distance L_{eff} using array theory. E_θ and E_ϕ can be written as [2]:

$$E_\theta = E_0 f(\theta,\phi)\cos\phi$$

(3.4)

$$E_\phi = -E_0 f(\theta,\phi)\cos\theta\sin\phi$$

(3.5)

where

$$f(\theta,\phi) = \frac{\sin\left[\dfrac{kW}{2}\sin\theta\sin\phi\right]}{\dfrac{kW}{2}\sin\theta\sin\phi}\cos\left(\dfrac{kL_{eff}}{2}\sin\theta\sin\phi\right)$$

(3.6)

k is the free-space wave number, and E_0 is the field strength maximum.

The E plane and H plane polar patterns are given by [2]:

$$F_E(\theta) = \cos\left[\frac{kL_{eff}}{2}\sin\theta\right]$$

(3.7)

$$F_H(\theta) = \cos\theta \, \dfrac{\sin\left[\dfrac{kW}{2}\sin\theta\right]}{\dfrac{kW}{2}\sin\theta} \tag{3.8}$$

The E plane of radiation is the $\phi = 0°$ cut when the radiating edges of the patch lie parallel to the x axis in Figure 3.1(a), and the H plane is the $\phi = 90°$ cut for the same condition.

A square patch antenna with dimensions $L = W = 55$ mm, fabricated on duroid (relative permittivity $\varepsilon_r = 2.2$ and loss tangent 0.001) with thickness $t = 1.6$ mm has a fundamental resonant frequency of 1.795 GHz. The E and H plane radiation patterns at this frequency are given in Figures 3.2 and

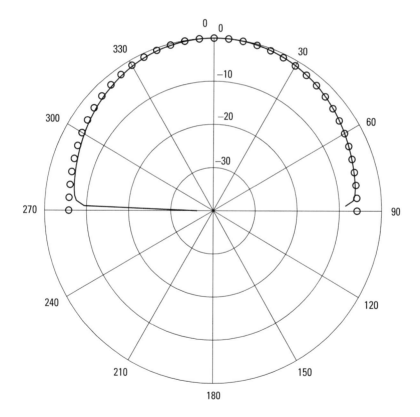

Figure 3.2 E plane radiation pattern (normalized dB) for a square patch antenna ($\phi = 90°$ cut); '-' Ensemble [7] result, 'o' (3.7) result.

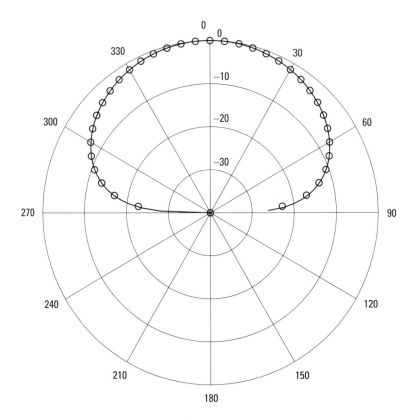

Figure 3.3 H plane radiation pattern (normalized dB) for a square patch antenna ($\phi = 0°$ cut); '-' Ensemble [7] result, 'o' (3.8) result.

3.3, respectively. It has been assumed that the feed point is at its optimum for coupling to a 50-Ω transmission line, and that the ground plane is of infinite extent in the xy plane. These patterns have been calculated using the method of moments (MOM) code, Ensemble [7], with the feed position ($x_0 = 27.5$ mm, $y_0 = 20.49$ mm). The normalized patterns from (3.7) and (3.8) have been included for comparison.

The maximum gain of the antenna lies in the $\theta = 0°$ direction (i.e., perpendicular to the patch surface). The E plane radiation pattern is almost omnidirectional, with nulls along the surface of the dielectric only. The agreement between the results obtained using the Ensemble [7] software and (3.7) and (3.8) is quite strong. In the E plane pattern, the Ensemble [7] result correctly shows a null at $\theta = 90°$, which is not included in (3.7).

Figure 3.4 illustrates the variation in the reflection coefficient of the same square patch for a number of different feed positions measured relative to one end. The data shown was obtained using the cavity model equations given in [3, 8], but is almost identical to that obtained using Ensemble [7]. The impedance bandwidth ($S_{11} < -10$ dB) is a maximum when the feed point is located at a distance of 19 mm from the edge; however, the best input match is obtained when the feed point is located 20 mm from the edge.

The input resistance R_A at the edge of a resonant rectangular patch is given by the approximate expression [9]:

$$R_A \cong 90 \frac{e_r}{pc_1} \varepsilon_r \mu_r \left(\frac{L}{W} \right)^2 \tag{3.9}$$

where e_r is the radiation efficiency of a horizontal electric dipole on top of the substrate, which is an approximation for the radiation efficiency of a single

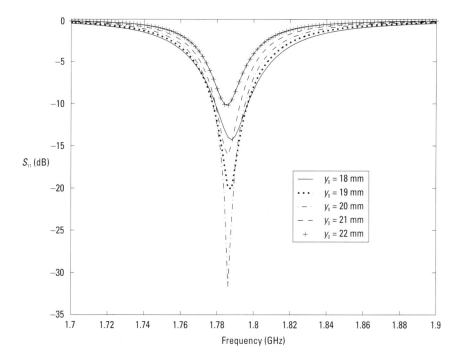

Figure 3.4 S_{11} (dB) of a coaxially fed square patch antenna plotted as a function of frequency for various feed positions.

patch, and p is referred to as the p factor and is the ratio of the power radiated into space by the patch to the power radiated into space by an equivalent dipole. The equation for p is given in [8].

$$c_1 = 1 - \frac{1}{\varepsilon_r \mu_r} + \frac{2}{5(\varepsilon_r \mu_r)^2} \qquad (3.10)$$

$$e_r = \frac{P_r^h}{P_r^h + P_{SW}^h} \qquad (3.11)$$

where P_r^h is the space-wave radiated power of a horizontal electric dipole on the substrate. The equation for P_r^h can be found in [8]. P_{SW}^h is the surface-wave radiated power of a horizontal electric dipole on the substrate. The equation for P_{SW}^h is also given in [8].

The equation for R_A gives typical input resistance values ranging from 100 to 400 Ω with the resistance reduced by increasing W. This can be useful, as there is no reactance at resonance and so the value obtained for input resistance should be quite close to that obtained using the much more complicated cavity model at the resonant frequency. This equation is dependent on frequency; however, the resistance value obtained does not change much with frequency, and the results are only accurate around the resonant frequency. As a result, it is necessary to already know the resonant frequency of the patch in order for the results to be useful.

The current distribution across the patch length L at resonance is a half-wavelength sinusoid, and the voltage distribution is a half-wavelength cosinusoidal function. The impedance varies significantly across the patch. For a coaxial probe, the distance between the probe and the edge of the patch (x_0) decreases the input resistance by the factor [2]:

$$\cos^2\left(\frac{\pi x_0}{L}\right) \qquad (3.12)$$

This factor is similar to that for the offset feed in a dipole wire antenna given by (2.33). Figure 3.5 shows a comparison between the input resistance obtained using the cavity model [8] and that obtained using (3.9), with the correction for the feed probe calculated using (3.12) as the feed position is moved from the edge of the patch to the center of the patch. Figure 3.6 shows a comparison between the same techniques as the width of the patch

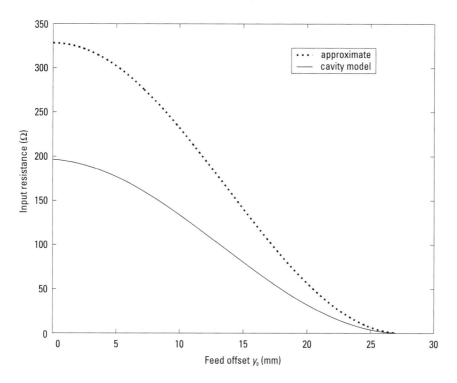

Figure 3.5 Input resistance (Ω) of a resonant square patch antenna obtained using the approximation of (3.9) and (3.12) compared with the cavity model as the feed position is moved from the edge of the patch to the center of the patch.

is varied from 30 mm to 80 mm with the feed offset (x_0) set constant at 20 mm.

The width of the patch is important for several reasons. First, the width affects the radiation pattern through (3.8) and the input resistance of the antenna through (3.9) (as shown in Figure 3.6). Figure 3.7 shows the variation in S_{11} as the width of the patch W is changed from 45 mm to 65 mm. The resonant frequency of the patch does not change significantly, which is expected because the resonant length of the patch remains constant. The impedance match is affected by changing W.

Second, a coaxial probe will excite currents in both the x and y directions. This may result in a large cross-polar component in the radiated field. This can be avoided by centering the feed probe along the width of the patch and offsetting the probe along the length of the patch.

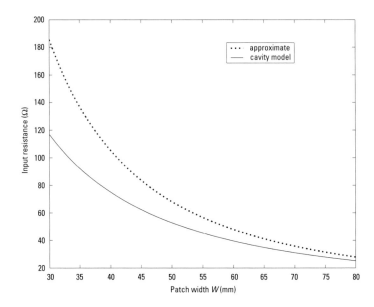

Figure 3.6 Input resistance (Ω) of a resonant square patch antenna obtained using the approximation of (3.9) and (3.12) compared with the cavity model as the width of the patch is varied from 30 mm to 80 mm.

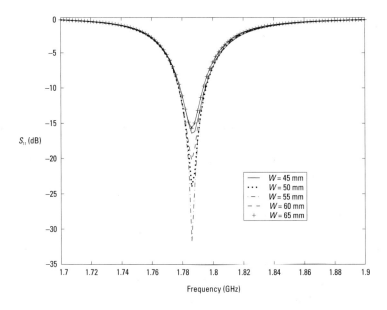

Figure 3.7 S_{11} (dB) of a coaxial feed patch as a function of frequency for various patch widths.

One disadvantage found in using the probe feed is that the patch will not be resonant when $t > 0.1\lambda_0$ because of the inductance introduced by the feed wire [2].

The bandwidth B and radiation efficiency e_r of a patch antenna both decrease as ε_r increases. The bandwidth increases as t increases, and the radiation efficiency decreases as t increases; hence, there is always a tradeoff between bandwidth and radiation efficiency, as a high radiation efficiency is required and a wide bandwidth is desirable. These two effects increase the sidelobe levels and cross-polar component in the radiation field. An approximation for the radiation efficiency was given in (3.11), and an approximation for the bandwidth is [9]:

$$B = 3.77 \frac{c_1 p}{e_r} \left(\frac{1}{\varepsilon_r} \right) \left(\frac{t}{\lambda_0} \right) \left(\frac{W}{L} \right) \tag{3.13}$$

where B is defined as the fractional bandwidth relative to the center frequency for a VSWR less than 2:1. This is equivalent to an S_{11} value of -9.5 dB.

The bandwidth and radiation efficiency of the square patch are plotted in Figures 3.8 and 3.9 versus the substrate height, for various values of substrate dielectric constant ($\varepsilon_r = 2.2, 4.4, 6.15$, and 10.2).

The radiation pattern of a patch antenna is dependent on the size of the ground plane [10]. Figures 3.2 and 3.3 were calculated on the assumption that the ground plane was infinite. These plots show no radiation at angles $|\theta| > 90°$. When the ground plane has finite size, diffraction into these angles results in a measurable front-to-back ratio. This can be calculated using the geometrical theory of diffraction [1, 10] or using a numerical solver such as Ensemble [7]. The ground plane should extend at least $3t$ past the edge of the patch, where t is the thickness of the dielectric. The square patch ($W = L = 55$ mm, $x_0 = 20$ mm, $y_0 = 27.5$ mm) has been modeled on Ensemble [7] with square ground planes, ranging in length from $0.57\lambda_0$ (equal to $6t + 55$ mm) to $1\lambda_0$. The E and H plane radiation patterns are shown in Figure 3.10 and 3.11 and the S_{11} is shown in Figure 3.12. As the ground plane size is increased, the radiation below the ground plane decreases, as expected, until for an infinite ground plane, there is no radiation below the ground plane.

Patch antennas can be designed to radiate with circular polarization. Left-hand circular polarization (LHCP) radiation is defined when the electric field vector rotates clockwise, and right-hand circular polarization (RHCP) when the electric field vector rotates counterclockwise. In limiting this

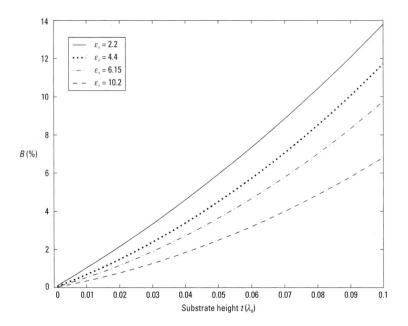

Figure 3.8 Bandwidth B (%) of a square patch antenna versus the substrate height for various values of the substrate dielectric constant.

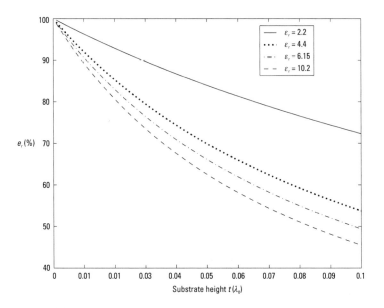

Figure 3.9 Radiation efficiency e_r (%) of a square patch antenna versus the substrate height for various values of the substrate dielectric constant.

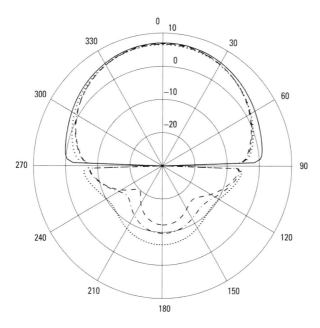

Figure 3.10 E plane radiation pattern (dBi) of a square patch antenna on various finite ground planes. '....' $0.57\lambda_0$, '_ . _' $0.78\lambda_0$, '_ _ _' $1\lambda_0$, '____' infinite.

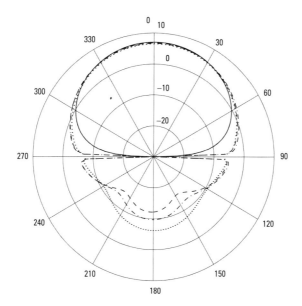

Figure 3.11 H plane radiation pattern (dBi) of a square patch antenna on various finite ground planes. '....' $0.57\lambda_0$, '_ . _' $0.78\lambda_0$, '_ _ _' $1\lambda_0$, '____' infinite.

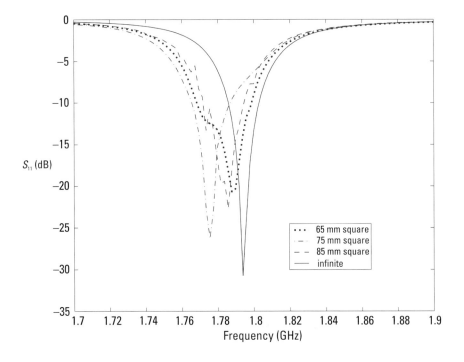

Figure 3.12 S_{11} (dB) of a square patch antenna for various finite size ground planes.

discussion to rectangular patches and feed methods in which the transmission lines are below the patch only, the viable methods of generating circular polarization are (see Figure 3.13):

1. Two simultaneous feeds for a square patch in which there is a 90° phase delay between them [Figure 3.13(a)];

2. Single feed, almost square patch, with element shaping, such as the removal of two opposite corners [5] [see Figure 3.13(b)] and/or the removal of some conductive material within the patch, such as a slot or cross structure [5] [Figure 3.13 (c)];

3. Single feed, almost square patch, with the feed placed in one corner of the patch [Figure 3.13(d)].

Square circularly polarized patches using the feed mechanisms shown in Figure 3.13 have been designed for operation around 1.79 GHz. For an isolated square patch with two feeds for circular polarization [Figure

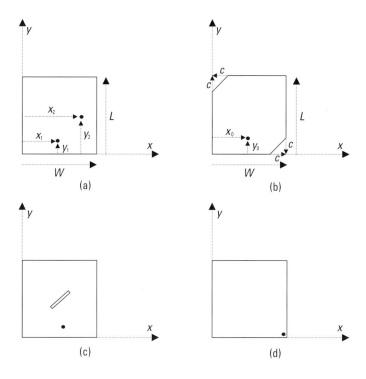

Figure 3.13 Circularly polarized rectangular patch antennas: (a) two simultaneous feeds for a square patch with 90° phase delay; and single feed, almost square patch with the removal of (b) two opposite corners or (c) a slot or cross structure, or with (d) the feed placed in one corner of the patch.

3.13(a)], the optimum design is obtained by placing the orthogonal feed with the same offset as the optimum linearly polarized feed. That is, for a 55-mm-by-55-mm square patch with $\varepsilon_r = 2.2$, feed 1 is placed at $x_1 = 27.5$ mm, $y_1 = 20$ mm, and feed 2 is placed at $x_2 = 35$ mm, $y_2 = 27.5$ mm. Figure 3.14 shows the reflection coefficient of both feeds, which are almost identical, and also shows the mutual coupling between feeds, which is minimal. Figure 3.15 shows the LHCP and RHCP radiation patterns in the $\phi = 0°$ plane and Figure 3.16 shows the same patterns in the $\phi = 90°$ plane. To generate these plots, feed 1 was assumed to have a 0° phase offset, and feed 2 a 90° phase offset relative to feed 1. The figures show that this generates an LHCP sense. If the phase offsets are reversed so that feed 1 is 90° relative to feed 2, the polarization would be RHCP. For a circularly polarized antenna with LHCP, the RHCP represents the cross-polar component, and vice versa.

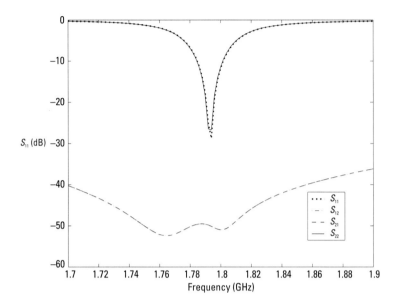

Figure 3.14 Scattering parameters (dB) of a circularly polarized square patch antenna with two feed positions.

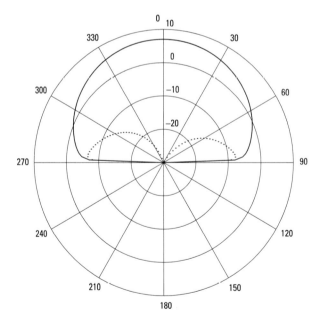

Figure 3.15 LHCP and RHCP radiation patterns (dBi) for the square circularly polarized patch in the $\phi = 0°$ plane at 1.794 GHz. '____' is LHCP; '_ _ _' is RHCP.

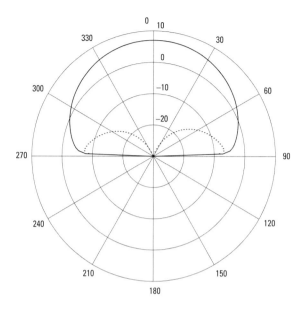

Figure 3.16 LHCP and RHCP radiation patterns (dBi) for the square circularly polarized patch in the $\phi = 90°$ plane at 1.794 GHz. '___' is LHCP; '_ _ _' is RHCP.

The truncated corner square patch ($L = W$) shown in Figure 3.13(b) generates circular polarization by producing two resonant frequencies due to the two different diagonal lengths, with a phase difference of 90° between the two. The choice of the length of the truncation, c, in Figure 3.13(b), is discussed in [10, 11]. It has been found that the optimum value for c is given by:

$$c = L\sqrt{\Delta s / S} \tag{3.14}$$

where $\Delta s / S$ is the truncation ratio given by:

$$\Delta s / S = \frac{1}{2Q_o} \tag{3.15}$$

and Q_o is the unloaded quality factor of the square patch, which can be calculated using the cavity model [3, 6].

For the 55-mm-by-55-mm square patch on a substrate with $t = 1.6$ mm and $\varepsilon_r = 2.2$, the optimum truncation length was calculated to be $c = 4.78$ mm. The position of the feed must be adjusted in order to obtain a good match for the circularly polarized case. The optimum feed position was found to be $x_0 = 27.5$ mm, $y_0 = 16$ mm. Figure 3.17 shows the reflection coefficient from Ensemble [7]. Figures 3.18 and 3.19 show the LHCP and RHCP radiation patterns for the $\phi = 0°$ and $\phi = 90°$ planes, respectively. This configuration produces LHCP, as seen by the radiation pattern plots. To produce RHCP, the position of the feed could be placed at $x_0 = 39$ mm, $y_0 = 27.5$ mm, or the opposite corners could be truncated.

3.2 Mutual Coupling Between Patch Elements

A dielectric slab on top of a perfectly conducting ground plane can act as a dielectric waveguide in which most of the field is confined in the dielectric material [12]. The mode of propagation is commonly referred to as a trapped surface-wave mode, and the field strength decreases exponentially with distance away from the dielectric surface. The rectangular patch antenna shown in Figure 3.1 provides an ideal launch for surface waves when the effective

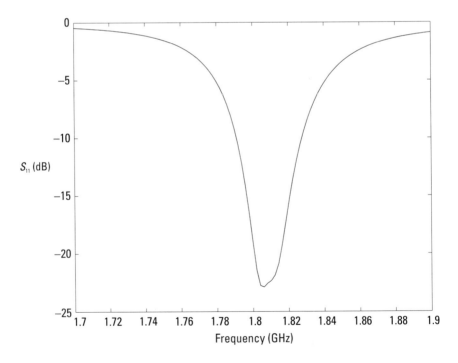

Figure 3.17 S_{11} (dB) of a square circularly polarized patch with truncated corners.

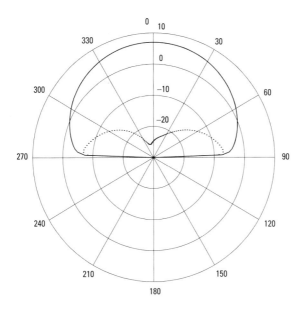

Figure 3.18 LHCP and RHCP radiation patterns (dBi) for a square circularly polarized patch with truncated corners in the ϕ = 0° plane at 1.806 GHz. '____' is LHCP; '_ _ _' is RHCP.

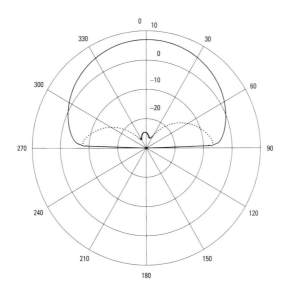

Figure 3.19 LHCP and RHCP radiation patterns (dBi) for a square circularly polarized patch with truncated corners in the ϕ = 90° plane at 1.806 GHz. '___' is LHCP; '_ _ _' is RHCP.

thickness of the dielectric material is sufficiently large, that is, when either t or ε_r or both are large. In some cases, surface waves may be the dominant coupling mechanism between patches and other conductors located on the same surface of the dielectric slab. Other coupling mechanisms include the space wave, leaky waves, and other higher-order waves [1].

For all patch antennas, it is desirable to minimize the energy transferred to surface waves, as it is desirable to radiate as much energy as possible. A grounded dielectric slab can contain TE and TM surface-wave modes [13]. The TM_0 surface-wave mode has a cutoff frequency of zero, and so is always present. The remaining TE_0, TM_1, TE_1 ... surface wave modes have cutoff frequencies f_i where:

$$f_i = \frac{c}{4t\sqrt{\varepsilon_r - 1}} \tag{3.16}$$

As a general guide to the design of patch antennas, the thickness of the dielectric t is chosen to minimize the surface-wave mode through the approximation [3]:

$$t \leq \frac{c}{4f\sqrt{\varepsilon_r - 1}} \tag{3.17}$$

This approximation ensures that only the TM_0 surface-wave mode can propagate, and so most of the energy is transferred to space waves.

In addition to mutual coupling, the effect of the surface wave is to decrease the radiation-field strength and to increase the sidelobe levels in patch antennas with ground planes of finite size. This is the result of the surface wave propagating to the end of the ground plane, where some energy is radiated.

When a conductive element lies close to the driven patch radiator, the current in the element contributes to the final radiation characteristics. If the conductive element has a resonant frequency close to the driven frequency, then the current in the parasitic element may be large. As with wire antenna elements, parasitic patch elements of a slightly larger size compared with a resonant element will act as reflectors, and those of slightly smaller size as directors [14]. It is also possible to add phasing elements (open-circuit microstrip lines or lumped elements) to patches to change their resonant frequency [5]. The net result is an increase in the bandwidth of the antenna [4] and a deviation of the main-beam direction from the surface normal.

Unlike the side-by-side configuration of wire antennas, if the parasitic patch elements are oriented so that the nonradiating edges are parallel, they lie in the radiation null, and so the coupling is not as strong as it is when the radiating edges are parallel.

There are two possible locations for the parasitic element. The E plane configuration is when the parasitic element is located next to the radiating edge of the driven patch, such that the two radiating edges are parallel [Figure 3.20(a)]. The H plane configuration is when the parasitic element is located next to the nonradiating edge of the driven patch, that is, the nonradiating edges are parallel [Figure 3.20(b)].

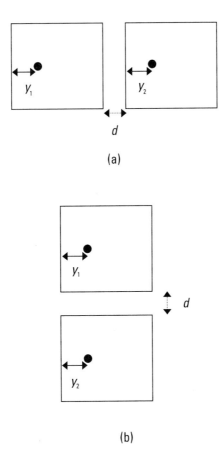

Figure 3.20 Two-element patch antenna array: (a) E plane configuration, and (b) H plane configuration.

The mutual impedance calculation for patch antennas is far more complicated in comparison to that for wire antennas given in (2.37). Expressions for the mutual impedance for both the E plane and H plane arrangements for coupled rectangular patches have been derived by a number of authors [15–18] and are based on a Green's function analysis.

When using the matrix equations to represent a patch antenna network, the empirical definitions for the matrix elements must be observed. Z_{11} and Z_{22} are defined in network theory as the impedance into port 1 or 2 when all other ports of the network are open-circuited. Similarly, Z_{12} and Z_{21} are the mutual impedance between ports 1 and 2 or 2 and 1 when all other ports are open-circuited.

In the wire array analysis using the network matrix equations, parasitic elements had their feed positions either open- or short-circuited. If the wire elements were resonant lengths, an open-circuited feed would result in two nonresonant wires, and these wires would have little effect on the array performance, that is, they would have very little induced current due to mutual coupling. As a result of this, the calculation of Z_{11} and Z_{22} are simplified by ignoring other elements of the array, and so these then become the self-impedances of the individual elements. Similarly, Z_{12} and Z_{21} are calculated in isolation and so are the mutual impedances between the two isolated elements. These approximations are valid when there is minimal induced current in the array elements with the feed positions open-circuited.

For the patch arrays, again the feed positions may be short- or open-circuited; however, with patch elements, there may still be a considerable induced current in a parasitic element even with the feed open-circuited. This will be dependent on the dimensions of the patches and the separation distance between the active patch and the parasitic patch. However, this means that any approximations made for the Z parameters will have an adverse effect on the accuracy of the network technique, and so all Z parameters should be calculated using their correct network definitions for accuracy. For this reason, Z_{11} and Z_{22} are calculated with all array elements present and their feeds open-circuited, and for patch arrays these are not the self-impedance of the patches. Similarly, Z_{12} and Z_{21} are calculated with all elements of the array present but with all other feeds open-circuited. This complicates the analysis of patch antenna arrays considerably. However, if the separation distance between the elements is large enough such that there is minimal induced current in an open-circuited parasitic element, then it may be possible to make the same approximations as for the wire arrays.

Ensemble [7] models were used to demonstrate this effect. Figure 3.21 shows Z_{11} of a square patch in isolation compared with Z_{11} for arrays of two and three patches. The arrays contain square patches of identical size ($L = W$ = 55 mm, t = 1.6 mm, and ε_r = 2.2), oriented for E plane or radiative edge coupling, and the separation distance between the patches is $d = 0.02\lambda_0$. (This separation distance is within the range required for switched parasitic arrays.) Figure 3.22 shows the Z_{11} for the same configurations but with H plane or nonradiative edge coupling.

Figures 3.23 and 3.24 show the Z_{11} for a single patch and arrays of two, three, and four patches versus the edge separation distance d between the patch edges for both E plane and H plane coupling. The Z_{11} for a single patch is constant and the Z_{11} for the arrays of two and three patches vary with the separation distance, but converge on the Z_{11} of the single patch for large separation distances ($>0.2\lambda_0$). The data was obtained at 1.795 GHz.

Figure 3.25 shows the mutual impedance, Z_{12}, for linear arrays of two, three, and four square patches for E plane coupling; Figure 3.26 shows the

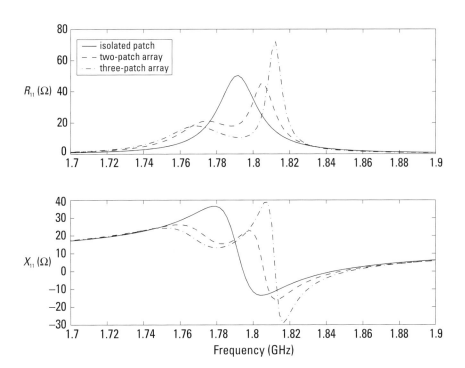

Figure 3.21 Z_{11} (Ω) for arrays of one, two, and three patches; the patches are oriented for E plane coupling and the separation distance between patch edges is $d = 0.02\lambda_0$.

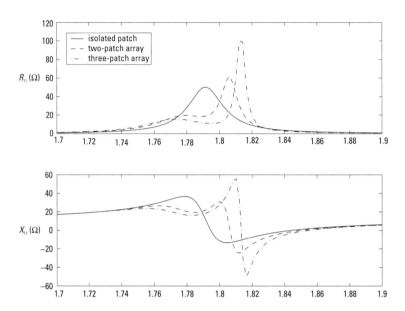

Figure 3.22 Z_{11} (Ω) for arrays of one, two, and three patches; the patches are oriented for H plane coupling and the separation distance between patch edges is $d = 0.02\lambda_0$.

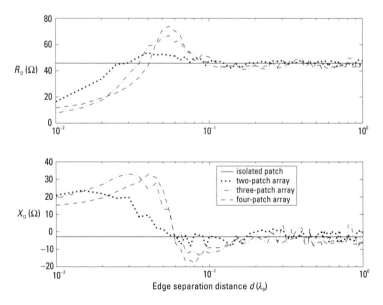

Figure 3.23 Z_{11} (Ω) at 1.795 GHz for arrays of one, two, three, and four patches versus the edge separation distance between the patches. The patches are oriented for E plane coupling.

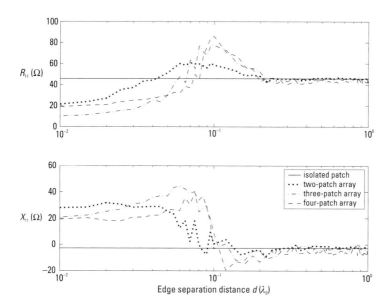

Figure 3.24 Z_{11} (Ω) at 1.795 GHz for arrays of one, two, three, and four patches versus the edge separation distance between the patches. The patches are oriented for H plane coupling.

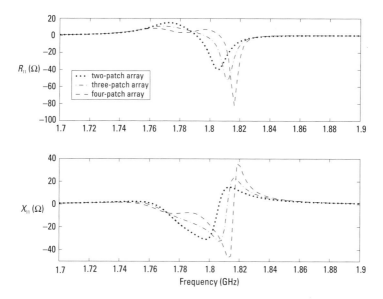

Figure 3.25 Z_{12} (Ω) for arrays of two, three, and four patches; the patches are oriented for E plane coupling and the separation distance between patch edges is $d = 0.02\lambda_0$.

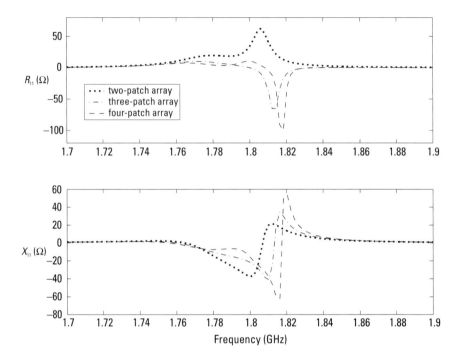

Figure 3.26 Z_{12} (Ω) for arrays of two, three, and four patches; the patches are oriented for H plane coupling and the separation distance between patch edges is $d = 0.02\lambda_0$.

same for H plane coupling. Figures 3.27 and 3.28 show the E and H plane Z_{12} for the same linear arrays, as the separation distance between the elements (d) is varied. These figures show that for both the E and H plane coupling cases, if d is greater than $0.2\lambda_0$, the two element array can be used to calculate the mutual impedance between all array elements.

The impedance bandwidth of patch antennas can be increased through the use of parasitic microstrip lines located close to the edges of the driven element [4]. These elements, however, do not result in a substantial change in the direction of the radiation from the array. A noticeable shift in beam direction occurs only if the parasitic elements have both x and y dimensions similar to those of a radiating patch antenna.

The effect of an adjacent parasitic patch located in the E plane (x axis) with a slightly higher resonant frequency is to pull the main radiation-beam direction slightly in this direction [14]. The direction of the maximum radiation is no longer at $\theta = 0°$. An adjacent patch with a slightly higher resonant

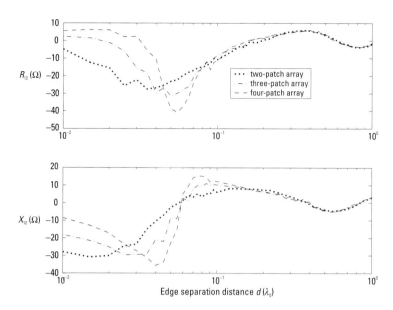

Figure 3.27 Z_{12} (Ω) at 1.795 GHz for arrays of two, three, and four patches versus the edge-separation distance between the patches. The patches are oriented for E plane coupling.

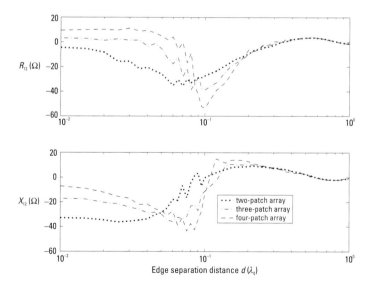

Figure 3.28 Z_{12} (Ω) at 1.795 GHz for arrays of two, three, and four patches versus the edge-separation distance between the patches. The patches are oriented for H plane coupling.

frequency shifts the main-beam direction away from the parasitic element. A combination of larger patches on one side of the driven element and smaller patches on the other side forms the basis of a Yagi-Uda patch antenna [14]. In these arrays, there is an increase in the bandwidth due to the slightly different resonant frequencies of the director and reflector patches. A four-element Yagi patch antenna array was designed as a conformal antenna with sufficient gain for ground-based communications to a geostationary satellite [14].

3.3 Patch Antenna Phased Arrays

The photolithographic process often used in the fabrication of patch antennas is ideally suited to the fabrication of arrays of patches. Given that the elements are arranged on the surface of a dielectric slab, it is straightforward to arrange the patches in a two-dimensional array. When the patches are sufficiently far apart and the substrate has been selected to minimize coupling between adjacent elements, the input impedance of each element is close to that for an isolated patch, and the radiation pattern can be determined from the array factor equations used for wire antennas. For example, three driven patches in the E plane configuration are shown in Figure 3.29. If all three patches are fed with the same voltage in phase, the primary beam direction will be at $\theta = 0°$, and the first null can be calculated from the equation:

$$\sin \theta = \frac{\lambda_0 / 2}{d + \lambda_g / 2} \qquad (3.18)$$

where d is the patch separation defined in Figure 3.29 and λ_g is the wavelength in the dielectric.

If the three patches are driven in the fundamental mode, then the E plane radiation pattern is given by [3]:

$$E_\theta = E_0 f(\theta, \phi) \cos \phi \sum_{i=1}^{3} I_i e^{-jkt_i \cos \theta} \qquad (3.19)$$

where E_0 is a scaling factor, $f(\theta, \phi)$ is given by (3.6), I_1, I_2, and I_3 are the currents in each of the patch elements, and t_i is the center-to-center separation distance between the elements. For a linear array, the H plane radiation pattern is identical to that for a single patch and is given by (3.8).

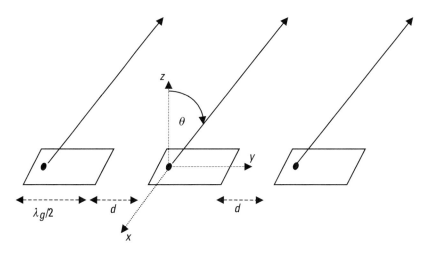

Figure 3.29 E plane array of half-wave patches, with the center-to-center separation distance of $(d + \lambda_g/2)$, where d is the nearest edge spacing.

In a two-dimensional array, both the E plane and H plane radiation patterns are altered, and the beam can be directed anywhere within the confines of the radiation pattern of an isolated patch.

In a phased array, every element in the array is active, and is fed with an RF signal that may be controlled in both magnitude and phase. The primary direction of the radiation and/or the location of the nulls in the radiation pattern can be steered electronically by controlling both the phase shifters and attenuators.

For beam scanning arrays, it is common to define an angular range over which there should be no grating lobes for optimal performance. This is used to determine the element spacing. In general, grating lobes occur when the center-to-center spacing of the antenna elements is greater than 1 (λ_0). If the spacing is less than 1 (λ_0), there will be no grating lobes, but there will be minor lobes. Depending on the angular range defined, the optimum center-to-center spacing may be less than or greater than 1 (λ_0).

3.4 Stacked Patch Antennas

The presence of a parasitic patch immediately above the driven patch (referred to as a stacked patch) can have two effects depending on the design. One effect is that a second resonance is found in the S_{11} characteristics resulting from the resonant frequency of the second patch. There is a limit to the frequency difference between the two resonances if the parasitic patch relies on proximity coupling from the driven patch in order for it to be resonant. The second effect is an increase in the bandwidth of the array [5]. This was discussed in Section 3.2. If the resonant frequency of the two patches is only slightly different, the antenna has effectively two overlapping resonances. This will have little change on the radiation pattern but changes the input impedance of the driven element significantly at the resonant frequency of the upper patch. Both techniques will increase the gain slightly due to the addition of a parasitic patch. Figure 3.30 shows the structure of a simple stacked patch antenna. Note that the dielectric material lying between the upper and lower patches can be different from that between the lower patch and the ground plane.

There are two techniques for feeding a probe-fed stacked patch. One is to have the probe feeding the lower element, in which case the upper patch is generally larger than or the same size as the lower patch [11]. The other involves feeding the upper patch through a cavity in the substrates and an access hole on the lower patch, in which case the upper patch is generally smaller than the lower patch, but may be slightly larger [11]. The upper probe case is generally used for dual-frequency designs, where a large separation between the two frequencies is required. The coupling to the lower patch in this case is from the probe passing through the hole in the substrate,

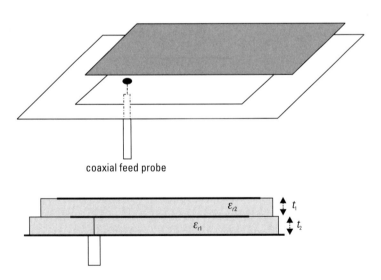

coaxial feed probe

Figure 3.30 Stacked patch antenna with a larger parasitic upper patch overlying the probe-fed lower patch.

and so both patches will be resonant and are not dependant on each other. The lower probe case is generally used for broadband patches. In this case, the coupling to the upper patch is caused by induced currents, and so the coupling is reduced as the difference between the resonant frequencies of the two patches increases.

Figure 3.31 shows the reflection coefficient versus frequency for a stacked patch designed to have a wide VSWR > 2 bandwidth. The lower patch is 55 mm square on a substrate with $\varepsilon_{r1} = 2.2$ and $t_1 = 1.6$ mm, and the upper patch is 60 mm square on a substrate with $\varepsilon_{r2} = 1.2$ and $t_2 = 11$ mm. The feed is offset by 0.5 mm from the edge of the patch (y_0). This figure shows a bandwidth of 12.45%. Figures 3.32 and 3.33 show the E and H plane radiation patterns at the center frequency and the band edges.

Figure 3.34 shows the reflection coefficient versus frequency for a stacked patch designed to have two separate resonant frequencies where the bottom patch is 55 mm square on a substrate with $\varepsilon_{r1} = 2.2$ and $t_1 = 1.6$ mm. The upper patch is 64 mm square on a substrate with $\varepsilon_{r2} = 1.2$ and $t_2 = 4.8$ mm. The feed offset for optimal impedance match is $y_0 = 9$ mm.

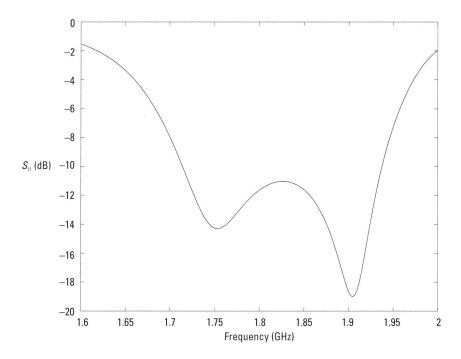

Figure 3.31 S_{11} (dB) for a broadband stacked patch with 12.45% bandwidth.

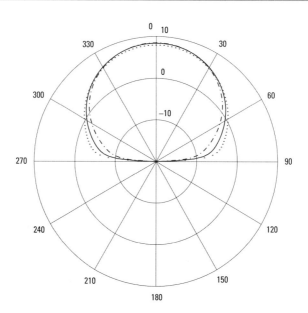

Figure 3.32 E plane radiation patterns (dBi) of the broadband stacked patch at the center frequency and the band edges. '....' 1.7147 GHz, '____' 1.8258 GHz, and '_ . _' 1.9369 GHz.

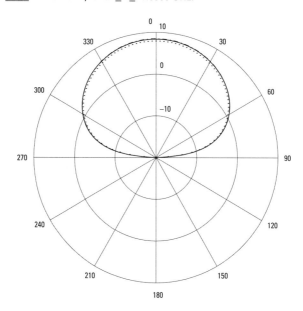

Figure 3.33 H plane radiation patterns (dBi) of the broadband stacked patch at the center frequency and the band edges. '....' 1.7147 GHz, '____' 1.8258 GHz, and '_ . _' 1.9369 GHz.

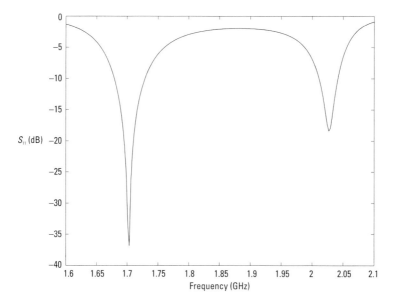

Figure 3.34 S_{11} (dB) for a dual-band stacked patch showing two separated resonant frequencies.

Figure 3.34 shows that there are two distinct resonant frequencies, 1.704 GHz and 2.027 GHz, with VSWR > 2 bandwidths of 3.1% and 1.6%, compared with the bandwidth of the single 55-mm square patch of 1.02%. Figures 3.35 and 3.36 show the E plane radiation patterns of the stacked patch at the two resonant frequencies, along with the patterns at the band edges. Figures 3.37 and 3.38 show the H plane patterns.

Dual-frequency directional antennas often have different radiation patterns at the two resonant frequencies. If one frequency is used for the uplink and the other is used simultaneously for the downlink, then there may be some uncertainty in determining the optimal direction of the antenna. It may be possible to design a parasitic array of stacked patches where the upper and lower patches are similar at the two resonant frequencies.

3.5 Switched Parasitic Patch Antennas

As with the wire antenna elements outlined in Section 2.7, it is possible to control the effect on the impedance of nearby parasitic patch elements and thereby change the direction of the main beam of a patch antenna.

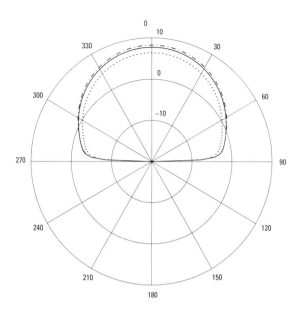

Figure 3.35 E plane radiation patterns (dBi) of the dual-band stacked patch at the lower resonant frequency and the band edges. '....' 1.678 GHz, '____' 1.704 GHz, and '_ . _' 1.730 GHz.

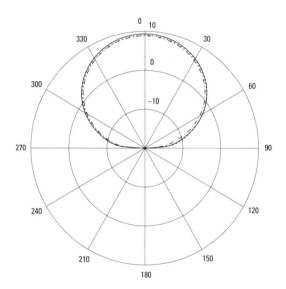

Figure 3.36 E plane radiation patterns (dBi) of the dual-band stacked patch at the upper resonant frequency and the band edges. '....' 2.011 GHz, '____' 2.027 GHz, and '_ . _' 2.043 GHz.

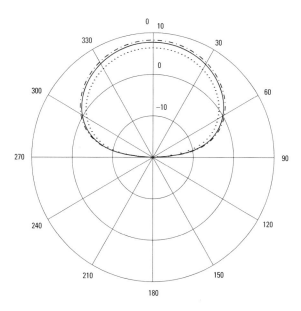

Figure 3.37 H plane radiation patterns (dBi) of the dual-band stacked patch at the lower resonant frequency and the band edges. '....' 1.678 GHz, '____' 1.704 GHz, and '_ . _' 1.730 GHz.

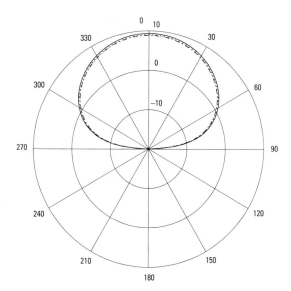

Figure 3.38 H plane radiation patterns (dBi) of the dual-band stacked patch at the upper resonant frequency and the band edges. '....' 2.011 GHz, '____' 2.027 GHz, and '_ . _' 2.043 GHz.

3.5.1   Switched Active Patch Arrays

The simplest switched parasitic patch array is shown in Figure 3.39. It consists of two patch elements oriented for maximum E plane coupling. One element is active and the other element has the feed position open-circuited. The position of the feed can be switched between either element, and so the beam can be directed in two positions separated by 180° in azimuth. This is similar to the wire SASPAs discussed in Chapter 2, and is referred to as a patch SASPA. The array consists of two 55-mm square patches, with feed offset ($y_1 = y_2$) of 14 mm constructed on a substrate with $\varepsilon_r = 2.2$ mm and $t = 1.6$ mm. The separation distance between the edge of the two elements is $d = 1.5$ mm. This separation distance equates to $0.009\lambda_0$ and gives the best directional radiation pattern.

Figure 3.40 shows the reflection coefficient obtained from Ensemble [7], with a VSWR < 2 bandwidth of 0.6% in switch position 1 and 1.1% in switch position 2. Figures 3.41 and 3.42 show the E and H plane radiation patterns also from Ensemble [7] in both switch positions. Position 1 is defined when element 1 in Figure 3.39 is open circuit and element 2 is active, and position 2 is when element 1 is active and element 2 is open circuit. The maximum gain of the beam in position 1 is 8.114 dBi and in position 2 is 8.306 dBi. In the E plane radiation plot, the change in signal strength between the two switch positions at $\theta = 30°$ and 60° is 8 dB and 15 dB respectively. As expected, there is almost no change in the H plane radiation pattern. For this particular design, the direction of the maximum radiation has been shifted significantly ($+18°$ and $-16°$). The addition of more parasitic elements either side of the current ones will pull the beam down further. This is demonstrated in Chapter 4.

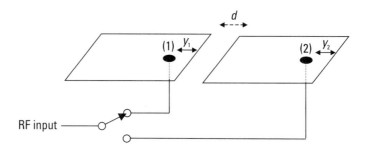

Figure 3.39 Two-element switched active patch array. The parasitic element feed position is always set at open circuit.

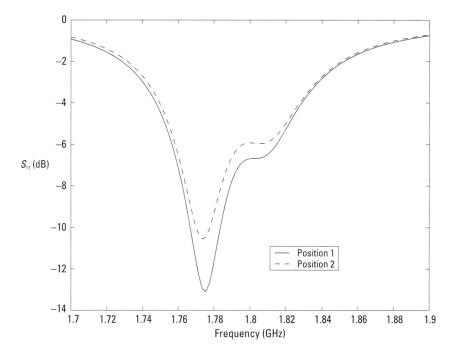

Figure 3.40 S_{11} (dB) of the two-element patch array shown in Figure 3.39 in both switch positions.

3.5.2 Switched Parasitic Patch Arrays

There are some advantages to fixing the position of the active element in an array, as described in relation to the wire arrays. These arrays are referred to as fixed active switched parasitic arrays (FASPA). A three-element switched parasitic patch array is shown in Figure 3.43. This switched parasitic patch array has a fixed active element located centrally, and a directional beam is formed when one of the outside elements is open-circuited and the other is short-circuited. The beam has two possible directions separated by 180° in azimuth, which are achieved by switching the position of the short- and open-circuited elements simultaneously. Figure 3.44 shows the reflection coefficient in both switch positions for an array of three 55-mm square patches with feed offset of 14 mm on a substrate with $\varepsilon_r = 2.2$ and $t = 1.6$ mm, and edge-to-edge separation of 1.5 mm. Position 1 is defined when element 1 is active, element 2 is open circuit, and element 3 is short-circuited. Position 2 is when element 1 is active, element 2 is short-circuited, and element 3 is open circuit. The VSWR < 2 bandwidth for position 1 is 1.2% and for position 2 is 1.4%. Figures 3.45 and 3.46 show the E and H plane radiation

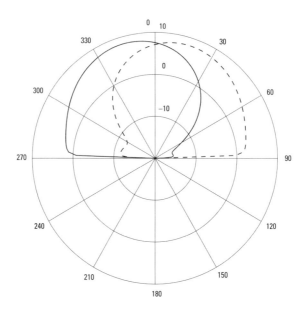

Figure 3.41 E plane radiation patterns (dBi) of the two-element patch array for both switch positions at 1.776 GHz. '____' Position 1; '_ _ _' position 2.

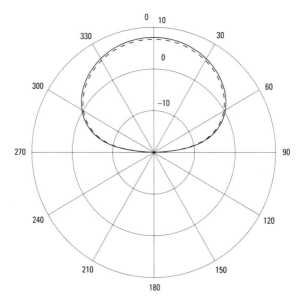

Figure 3.42 H plane radiation patterns (dBi) of the two-element patch array for both switch positions at 1.776 GHz. '____' Position 1; '_ _ _' position 2.

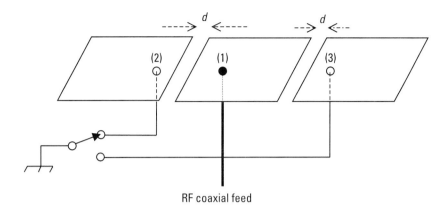

Figure 3.43 Three-element switched parasitic patch array.

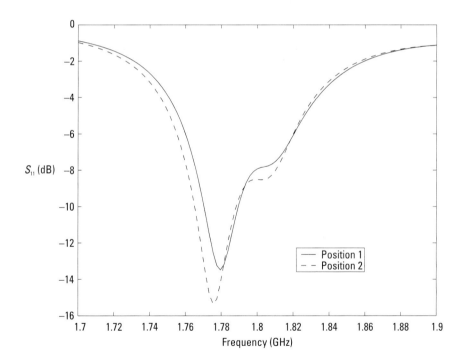

Figure 3.44 S_{11} (dB) of the three-element patch array shown in Figure 3.43 in both switch positions.

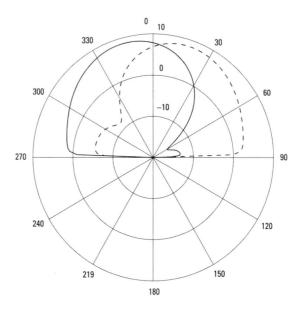

Figure 3.45 E plane radiation patterns (dBi) of the three-element patch array for the two parasitic element switch cases at 1.78 GHz.

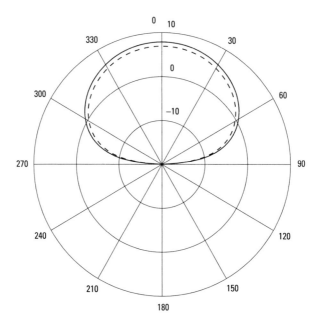

Figure 3.46 H plane radiation patterns (dBi) of the three-element patch array for the two parasitic element switch cases at 1.78 GHz.

patterns for the array in both switch positions. The maximum gain in position 1 is 8.619 dBi at $\theta = -14°$ and in position 2 is 8.393 dBi at $\theta = 20°$.

3.5.3 Circularly Polarized Switched Parasitic Patch Arrays

Switched parasitic circularly polarized patch antennas can be designed from square patches using an orthogonal feed probe, as described in Section 3.1. There is a slight difference in this case, however, because the two feed probes may be offset by different amounts due to the presence of the parasitic elements in only one plane (generally the E plane). Figure 3.47 shows a three-element fixed active switched parasitic circularly polarized patch array. The array elements are all 55-mm square, the edge separation distance between the central element and both parasitic elements is $d = 1.5$ mm, feed probe 1 is offset by 13 mm ($y_1 = 13$ mm), and 2 is offset by 12 mm ($x_2 = 43$ mm). The substrate has $\varepsilon_r = 2.2$ and $t = 1.6$ mm.

Figure 3.48 shows the reflection coefficient of each feed obtained from Ensemble [7] for both parasitic switch positions. The coupling between feeds for this array configuration was below −25 dB for both switch positions. Figures 3.49 and 3.50 show the LHCP radiation patterns in both switch positions in the $\phi = 0°$ and $\phi = 90°$ planes, respectively. To generate LHCP, feed 1 was assumed to have a 0° phase offset and feed 2 a 90° phase offset relative to feed 1. If the phase offsets are reversed, RHCP will be generated. In the figures, position 1 refers to the case when probe 3 is open circuit and 4 is short-circuited, and position 2 refers to the case when probe 3 is short-circuited and 4 is open circuit.

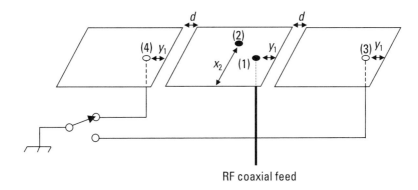

Figure 3.47 Three-element circularly polarized fixed active switched parasitic patch array.

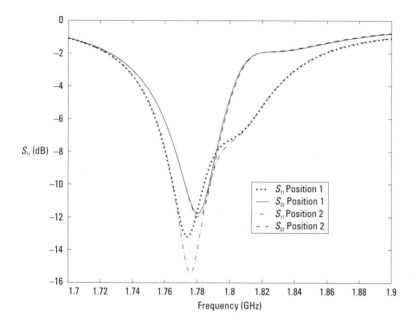

Figure 3.48 S_{11} (dB) of the three-element circularly polarized patch array shown in Figure 3.47.

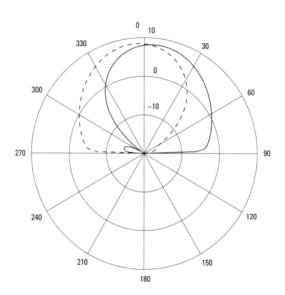

Figure 3.49 LHCP radiation patterns (dBi) for the three-element circularly polarized array in Figure 3.47 in the $\phi = 0°$ plane for both parasitic switch positions. '____' position 1; and '_ _ _' position 2.

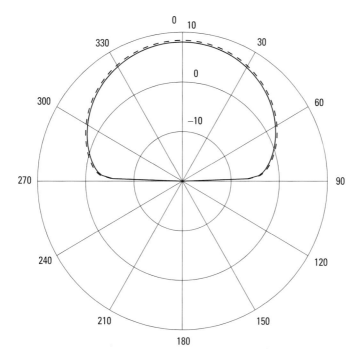

Figure 3.50 LHCP radiation patterns (dBi) for the three-element circularly polarized array in Figure 3.47 in the ϕ = 90° plane for both parasitic switch positions. '____' position 1; and '_ _ _' position 2.

The maximum gain for this array in position 1 is 8.24 dBi at $\theta = 8°$, and in position 2 is 8.46 dBi at $\theta = -6°$.

With the circularly polarized switched parasitic patch array, it may be useful to add parasitic elements adjacent to the orthogonal feed plane to achieve four beam directions separated by 90° in the ϕ plane, as shown in Figure 3.51. With all elements 55-mm square, all feeds offset by 12 mm, and an edge separation distance between elements of 1.5 mm, the reflection coefficient obtained using Ensemble [7] is shown in Figure 3.52 for all switch positions. Elements 1 and 2 are always active. Table 3.1 lists the four beam positions, with the switch configurations required for each.

Optimum LHCP is achieved when element 1 is fed at 0° phase and element 2 is fed at 70° with respect to element 1. Figure 3.53 shows the LHCP radiation patterns in the θ plane when $\phi = 0°$ for all switch positions. Figure 3.54 shows the LHCP radiation patterns when $\phi = 90°$ for all switch positions.

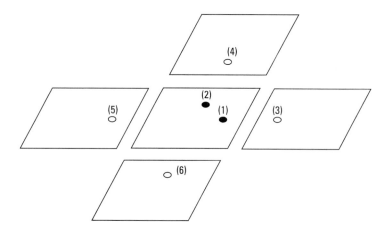

Figure 3.51 Five-element circularly polarized fixed active switched parasitic patch array.

3.6 Summary

As with the wire antenna case outlined in Chapter 2, patch antennas can be used as the basic array elements in switched parasitic and switched active antennas. These structures offer the same advantages as switched parasitic wire antennas, including a smaller footprint compared with phased arrays and lower insertion loss, as there are fewer RF feed lines involved in the construction. The beam can be switched from the vertical position (i.e., normal to the patch) in both directions when parasitic elements are located on all four sides of an active patch. Both linear and circular polarization is possible with square patch elements. The adjacent elements in switched parasitic antennas should be quite close together, whereas in phased arrays, a major problem occurs when the mutual coupling between elements is too great.

Table 3.1
Switch Configurations for the Patch Array Shown in Figure 3.51

	Position 1	Position 2	Position 3	Position 4
Short-circuit elements	4, 5, 6	3, 5, 6	3, 4, 6	3, 4, 5
Open-circuit element	3	4	5	6

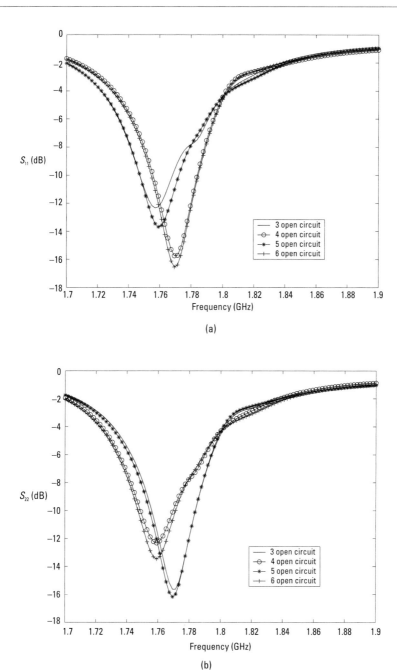

Figure 3.52 (a) S_{11} and (b) S_{22} (dB) of the switched parasitic patch array in Figure 3.51 in all four beam positions.

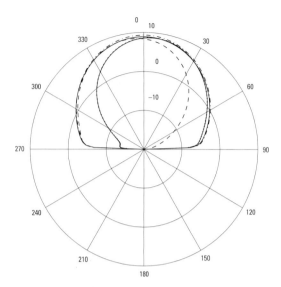

Figure 3.53 LHCP radiation patterns (dBi) of the circularly polarized switched parasitic patch array shown in Figure 3.51 in all switch positions in the θ plane when $\phi = 0°$. '____' position 1 (3 o/c), '_ _ _' position 2 (4 o/c), '_ . _' position 3 (5 o/c), and '....' position 4 (6 o/c).

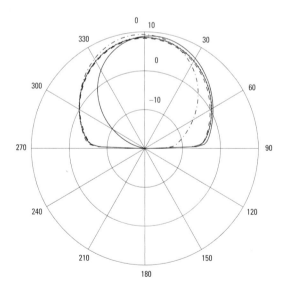

Figure 3.54 LHCP radiation patterns (dBi) of the circularly polarized switched parasitic patch array shown in Figure 3.51 in all switch positions in the θ plane when $\phi = 90°$. '____' position 1(3 o/c), '_ _ _' position 2 (4 o/c), '_ . _' position 3 (5 o/c), and '....' position 4 (6 o/c).

The mutual coupling between patch antennas cannot be described by a simple set of equations, and so it is necessary to use numerical methods to verify designs. In the next chapter, we will look at the design of more complex switched parasitic antennas.

References

[1] Balanis, C. A., *Antenna Theory Analysis and Design*, 2nd ed., New York: John Wiley and Sons, 1997.

[2] Stutzman, W. L., and G. A. Thiele, *Antenna Theory and Design,* 2nd ed., New York: John Wiley and Sons, 1998.

[3] Sainati, R. A., *CAD of Microstrip Antennas for Wireless Applications*, Norwood, MA: Artech House, 1996.

[4] Zurcher, J. F., and F. E. Gardiol, *Broadband Patch Antennas*, Norwood, MA: Artech House, 1995.

[5] Hirasawa, K., and M. Haneishi, *Analysis, Design, and Measurement of Small and Low-Profile Antennas*, Norwood, MA: Artech House, 1991.

[6] Gupta, K. C., and P. S. Hall (eds.), *Analysis and Design of Integrated Circuit Antenna Modules,* New York: John Wiley and Sons, 2000.

[7] Ansoft Coporation, Ensemble, 2000, http://www.ansoft.com.

[8] Jackson, D. R., and N. G. Alexopoulos, "Simple Approximate Formulas for Input Resistance, Bandwidth and Efficiency of a Rectangular Patch," *IEEE Trans. Antennas and Propagation*, Vol. 3, March 1991, pp. 407–410.

[9] Huang, J., "The Finite Ground Plane Effect on the Microstrip Antenna Radiation Pattern," *IEEE Trans. Antennas and Propagation*, Vol. 31, July 1983, pp. 649–653.

[10] Lee, K. F., and W. Chen, *Advances in Microstrip and Printed Antennas*, New York: John Wiley and Sons, 1997.

[11] James, J. R., and P. S. Hall, *Handbook of Microstrip Antennas*, London: Peter Peregrinus, 1989.

[12] Collins, R. E., *Field Theory of Guided Waves*, 2nd ed., New York: IEEE Press, 1991.

[13] Bhatia, P., K. V. S. Rao, and R. S. Tomar, *Millimeter-Wave Microstrip and Printed Circuit Antennas*, Norwood, MA: Artech House, 1991.

[14] Huang, J., "Planar Microstrip Yagi Array Antenna," *IEEE APS International Symposium Digest*, San José, CA, 1989, pp. 894–897.

[15] Derneryd, A. G., "A Theoretical Investigation of the Rectangular Microstrip Antenna Element," *IEEE Trans. Antennas and Propagation,* Vol. 27, No. 2, March 1979, pp. 137–145.

[16] Pozar, D. M., "Input Impedance and Mutual Coupling of Rectangular Microstrip Antennas," *IEEE Trans. Antennas and Propagation,* Vol. 30, No. 6, 1982, pp. 1191–1196.

[17] Diaz, L., and T. Milligan, *Antenna Engineering Using Physical Optics,* Norwood, MA: Artech House, 1996.

[18] Lee, K. F., and W. Chen, *Advances in Microstrip and Printed Antennas,* New York: John Wiley and Sons, 1997.

4

Design Examples of Switched Parasitic Antennas

4.1 Introduction

Parasitic elements have been used in antenna arrays since the 1920s, predominantly as directors and reflectors in Yagi-Uda arrays. The difference between a reflector parasitic element and a director parasitic element in a Yagi-Uda array is length. Researchers have since used a switch to change the effective length of a parasitic element, which can result in a change in the direction of radiation. Beam steering is possible when the switch is controlled electronically. The operation of an electronic switch is much faster than that achievable by a mechanically steered array.

In Chapters 2 and 3, the basic approach to switched parasitic wire and patch antennas was illustrated by simple two- and three-element arrays. Switched parasitic arrays with a larger number of elements can provide increased gain and smaller beamwidths. Many of the examples given here are much larger arrays that display this improved antenna performance.

The chapter is partly a historical survey of switched parasitic antennas, together with their basic properties. While it is impossible to describe all achievements in the field, some original designs are given, together with a number of the more unusual applications of the technology. Where possible, measured results are given for the arrays discussed. For cases where measured results are not available, modeled results are given. Unless otherwise stated, all wire arrays have been modeled on NEC [1] and all patch arrays have been modeled on Ensemble [2].

4.2 Historical Survey

4.2.1 Wire Arrays

Possibly the first use of switched parasitic elements in antenna arrays for direction finding was reported in 1971 by Himmel et al. [3]. This patent describes two monopole arrays, one for single-frequency operation and the other for dual-frequency operation. Both arrays consist of a single central active element surrounded by concentric rings of parasitic elements of various radii. The parasitic elements are either open-circuited or short-circuited to the ground plane via a diode switch. A directional radiation pattern is achieved by selecting the parasitic element switch states, and is rotated by sequentially changing the switch settings of the parasitic elements around the array.

Black et al. [4] patented a direction-finding array for aircraft navigation that used switched parasitic monopole elements on a ground plane formed by the aircraft fuselage. A diagram of the array is shown in Figure 4.1. A central active element (1) is surrounded by four parasitic elements (2–5). Each parasitic element is connected to the ground plane via a semiconductor switch. Parasitic elements 2 and 3 are slightly shorter than the active element, and elements 4 and 5 are slightly longer than the active element. This array produces two overlapping beams for direction finding. A directional beam is generated by shorting elements 2 and 4 to ground and by open-circuiting elements 3 and 5. In this case, element 2 is a director and element 4 a reflector. Similarly, shorting elements 3 and 5 to ground and open-circuiting elements 2 and 4 forms a directional beam pointing 100° from the first in the H plane.

The array shown in Figure 4.1 has the dimensions:

L_1 = 201.17 mm

L_2 = L_3 = 157.48 mm

L_4 = L_5 = 223.52 mm

D_1 = D_2 = 88.9 mm

D_3 = D_4 = 120.65 mm

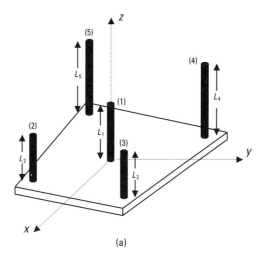

(a)

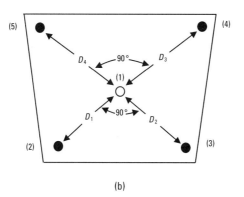

(b)

Figure 4.1 Aircraft direction-finding array [4]: (a) three-dimensional and (b) top views of array.

The antenna was modeled with an infinite ground plane and a monopole wire diameter of 6.35 mm. The S_{11} and H plane radiation patterns for both switch positions are shown in Figures 4.2 and 4.3, respectively. The radiation patterns were calculated at the resonant frequency of the array, f_0 = 350 MHz. The antenna has a main beam in the direction ϕ_{max} = 54° or ϕ_{max} = 306°, with a maximum gain of 6.45 dBi, a front-to-back ratio of 14.3 dB, a beamwidth B_{ϕ} = 134°, and an impedance bandwidth of 116 MHz or 33%.

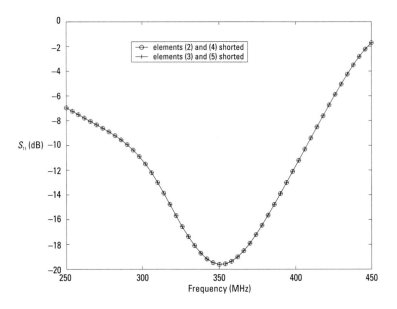

Figure 4.2 Frequency variation in S_{11} (dB) of the direction-finder array shown in Figure 4.1.

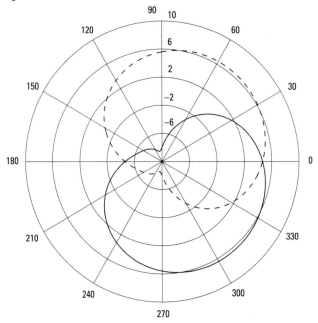

Figure 4.3 H plane radiation pattern (dBi) of the direction-finder array shown in Figure 4.1 in both switch positions.

A variation of an electronically steered Yagi-Uda array was given by Gueguen [5]. This monopole array was constructed on a ground plane with a central active element surrounded by rings of parasitic elements. The parasitic elements have varying lengths, so that when short-circuited to the ground plane, some form directors and some form reflectors. The parasitic elements are switched using *p.i.n.* diodes connected to the ground plane. A parabolic-shaped reflector formed by a number of short-circuited elements is arranged behind the active element.

The techniques presented so far have controlled the parasitic elements through the use of two switching states only, open- and short-circuited. Harrington [6, 7] investigated the possibility of using more than two switching states on the parasitic elements, and developed a reactively controlled directive array. A diagram of the seven-element dipole array described in [7] is shown in Figure 4.4. The array properties and the reactive loads required to give a number of beam directions were determined using a numerical optimization technique. The array consists of an active dipole element (1) and six surrounding parasitic dipole elements (2–7), spaced equally around a circle, with the active element at the center of the circle. All elements are $\lambda_0/2$ long; the radius defining the position of the parasitic elements for this example is $r = 0.2387\lambda_0$ and the radius of the wire dipole elements is $0.001\lambda_0$.

Table 4.1 gives the reactive loads required on each parasitic element to produce the beam directions listed. The modeled S_{11} and H plane radiation patterns for these beam directions are shown in Figures 4.5 and 4.6 respectively. The array was designed for operation at 300 MHz; however, the radiation patterns were calculated at 295 MHz.

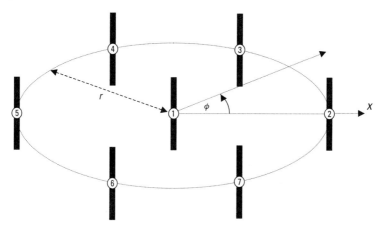

Figure 4.4 Reactively controlled array of dipoles designed by Harrington [6, 7].

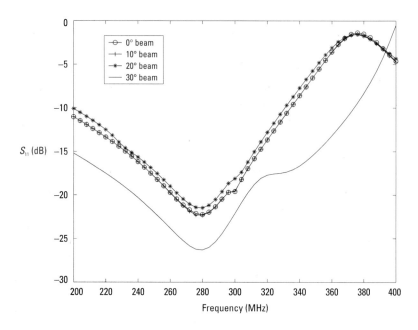

Figure 4.5 Frequency variation in S_{11} (dB) of the reactively controlled array shown in Figure 4.4 for the switch positions given in Table 4.1.

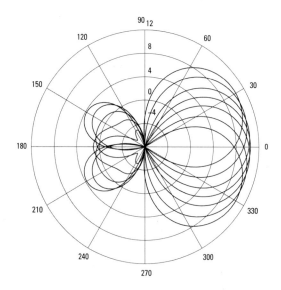

Figure 4.6 H plane radiation pattern (dBi), of the reactively controlled array shown in Figure 4.4 for the switch positions given in Table 4.1 at 295 MHz.

Figure 4.6 shows seven of the possible beams that can be generated by using the 22 different reactance values given in Table 4.1. The seven beams scan through a ϕ range of 60°. From the array symmetry, the beams shown can be scanned to cover all possible ϕ angles by rotating the reactance values around the array. The gain over 360° ranges from 10.39 dBi to 10.51 dBi with front-to-back ratios in the range 6.6 dB to 27 dB. The beamwidth of all beams is approximately 60°, and the impedance bandwidth is always greater than 130 MHz or 47%. Luzwick and Harrington published a similar technique using reactively loaded aperture elements [8]. The array consists of a number of waveguide-backed rectangular slots with the central element driven and the remaining elements short-circuited at a distance determined to give the required reactive loading.

In 1986 Dumas et al. [9] patented an array for direction finding and navigation similar to those described earlier in this section. This array consists of a central active element surrounded by a number of equally spaced parasitic elements. The parasitic elements are either capacitively or inductively coupled to ground, where a capacitively coupled parasitic element is a director and an inductively coupled element is a reflector. The parasitic elements are connected to a transmission line with an effective length controlled by a diode switch.

Milne [10, 11] outlines a switched parasitic array capable of steering in the elevation plane as well as the azimuth plane. The simplest array consists of a single active monopole element surrounded by two concentric rings of

Table 4.1
Reactive Loads Required on Each Parasitic Element to Generate a Beam in the Given Direction

Beam Direction $\phi_{max}(°)$	Reactive Loading (Ω)					
	(2)	(3)	(4)	(5)	(6)	(7)
−30	−57.4	−401.0	1.7	1.7	−398.1	−57.4
−20	−64.7	−134.1	−4.1	9.7	180.0	−71.0
−10	−64.3	−105.6	−1.0	11.4	34.8	−85.2
0	−63.6	−94.6	9.0	10.8	9.0	−94.6
10	−64.3	−85.2	34.8	11.4	−1.0	−105.6
20	−64.7	−71.0	180.0	9.7	−4.1	−134.1
30	−57.4	−57.4	−398.1	1.7	1.7	−401.0

parasitic elements. This array is capable of producing 16 beams separated by 22.5° in azimuth at $\theta = 90°$ and eight beams separated by 45° in azimuth at $\theta = 45°$ (with an infinite ground plane). In total there are 16 parasitic elements, eight on each ring. The parasitic elements are connected to the ground plane via $p.i.n.$ diodes. When a $p.i.n.$ diode is conducting, the parasitic element is a reflector; when it is not conducting, the parasitic element is virtually transparent in the array. A schematic diagram of the 17-element array is given in Figure 4.7. The concept was extended to larger arrays that give higher gain compared with that obtained from the 17-element array [11].

There are 24 possible beam directions with the 17-element array. Consider the case where the active element length $L_1 = 0.25\lambda_0$ and the parasitic element length $L_2 = 0.24\lambda_0$. The radius of the first ring of parasitic elements is $r_1 = 0.33\lambda_0$ and the radius of the second ring of parasitic elements is $r_2 = 0.5\lambda_0$. The array was modeled with a wire radius for all elements of $0.005\lambda_0$ and an infinite ground plane. The center element (1) is the active element. All parasitic elements listed in Table 4.2 are set to short circuit and all other elements are open circuit. Three different beam directions illustrate both the beam forming and the performance.

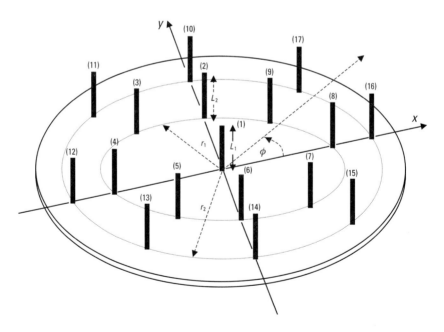

Figure 4.7 Seventeen-element adaptive array that provides scanning in both the θ and ϕ directions [10, 11].

Table 4.2
Beam Positions and Array Switch Settings for Three Beam Directions of the 17-Element Array Shown in Figure 4.7

	Direction	Short-Circuit Element Numbers	Beamwidth	Bandwidth
Direction 1	$\theta = 90°$ $\phi = 0°$	4, 10, 11, 12, 13, 14	66°	196 MHz/74%
Direction 2	$\theta = 45°$ $\phi = 0°$	10, 11, 12, 13, 14, 15, 17	84°	188 MHz/72%
Direction 3	$\theta = 90°$ $\phi = 22.5°$	4, 5, 10, 11, 12, 13, 14, 15	79°	208 MHz/79%

The modeled array was designed for operation at 300 MHz. The S_{11} and E and H plane radiation patterns at 300 MHz for these three beam directions are given in Figures 4.8 and 4.9, respectively. The gain of the beam in direction 1 is 9.12 dBi with a front-to-back ratio of 14.6 dB, the beam in direction 2 has a maximum gain of 4.67 dBi, and the beam in direction 3 has a gain of 8.62 dBi with a front-to-back ratio of 12.7 dB.

The E and H plane radiation patterns obtained at $\phi = n45°$ ($n = 1, 2, \ldots, 8$) for both the low- and high-elevation beams are identical. The radiation patterns obtained for $\phi = 22.5° + m45°$ ($m = 0, 1, \ldots, 7$) are also identical but different from those obtained at $n45°$ angles. For the beams generated at $\phi = 22.5° + m45°$, only the low-elevation beam is possible.

Recent research has been conducted on arrays of switched parasitic elements with more than one active element for application in wireless local area networks [12–14]. These papers discuss a circular array with two active elements and the remaining elements short-circuited to the ground plane. The beam is steered by switching the position of the active and parasitic elements around the array. For the array with two active elements, an n element array has a maximum of n beam directions. Should only one active element be used, the number of beam directions is doubled at the expense of antenna gain. It can be shown that with an array like this, it is possible to generate at least $2n$ beams by using a combination of between one and two active elements. In the published descriptions of this antenna [12–14], the parasitic elements are always short-circuited. A diagram of a six-element monopole array is shown in Figure 4.10.

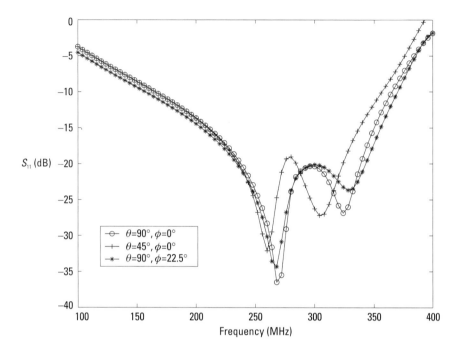

Figure 4.8 S_{11} (dB) of the 17-element Milne array for the three beam directions described in Table 4.2.

The six-element array shown in Figure 4.10 was modeled with an infinite ground plane. The element length $L = 0.25\lambda_0$, the radius defining the placement of the elements is $r = 0.25\lambda_0$, and the radius of the wires is $0.0125\lambda_0$. Table 4.3 gives the switch combinations required to generate 12 different beams, including those with both one and two active elements. It is possible to generate more beams than shown here by using a combination of open- and short-circuited parasitic elements.

Figures 4.11 and 4.12 show the S_{11} and H plane radiation patterns obtained at two of the possible switch locations (30° and 60°). The maximum gain of the array with a single active element ($\phi_{max} = 30°$) is 7.71 dBi, with a front-to-back ratio of 18.4 dB and a 3-dB beamwidth of 86°. The impedance bandwidth is 76%. The dual active element beam ($\phi_{max} = 60°$) has a maximum gain of 8 dBi, a front-to-back ratio of 13.6 dB, a 3-dB beamwidth of 78°, and an impedance bandwidth of 78%.

Figure 4.12 shows only two radiation patterns. The remaining radiation patterns for all other switch positions are identical when rotated through the appropriate angle. For example, from Table 4.3, $\phi_{max} = 30°$ is generated

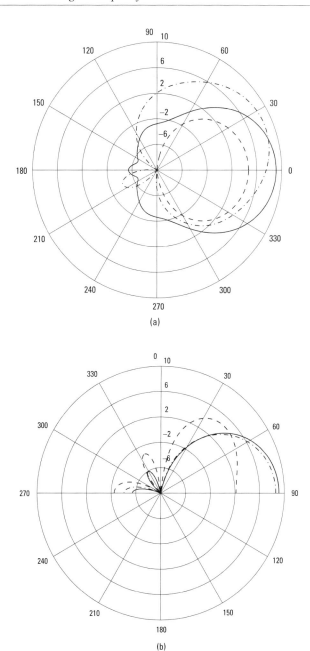

Figure 4.9 (a) H plane radiation patterns (dBi) of the 17-element Milne array at 300 MHz. '___' direction 1; '_ _ _' direction 2; and '_ . _' direction 3. (b) E plane radiation patterns (dBi) of the 17-element Milne array at 300 MHz. '___' direction 1; '_ _ _' direction 2; and '_ . _' direction 3.

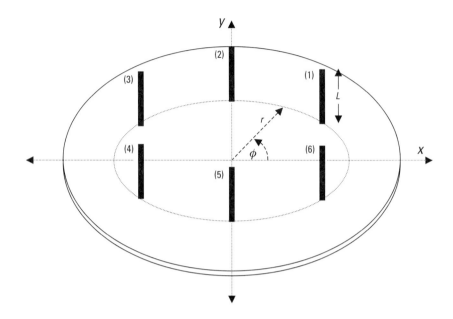

Figure 4.10 Six-element circular switched monopole array (Sibille et al. [12–14]).

Table 4.3
Switch Combinations Required to Generate 12 Different Beam Directions
for the Antenna in Figure 4.10

$\phi_{max}(°)$	Active Element/s	Short-Circuit Elements
0	1 and 6	2, 3, 4, 5
30	1	2, 3, 4, 5, 6
60	1 and 2	3, 4, 5, 6
90	2	1, 3, 4, 5, 6
120	2 and 3	1, 4, 5, 6
150	3	1, 2, 4, 5, 6
180	3 and 4	1, 2, 5, 6
210	4	1, 2, 3, 5, 6
240	4 and 5	1, 2, 3, 6
270	5	1, 2, 3, 4, 6
300	5 and 6	1, 2, 3, 4
330	6	1, 2, 3, 4, 5

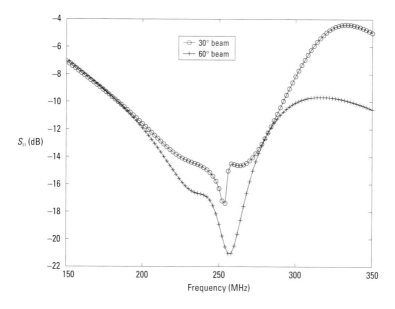

Figure 4.11 S_{11} (dB) of the array shown in Figure 4.10 in the two beam switch positions: $\phi_{max} = 30°$ and $\phi_{max} = 60°$.

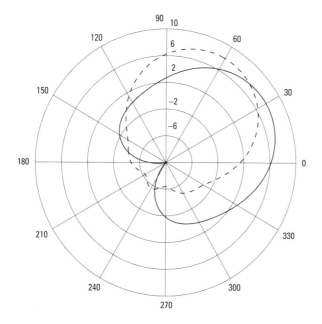

Figure 4.12 H plane radiation patterns (dBi) of the array shown in Figure 4.10 in the two beam switch positions: '____' $\phi_{max} = 30°$, and '_ _ _' $\phi_{max} = 60°$.

when element 1 is active and the remaining elements are short-circuited to ground. Identical radiation patterns are generated at $\phi_{max} = 90°$, $150°$, $210°$, $270°$, and $330°$. The $\phi_{max} = 60°$ radiation pattern is generated when elements 1 and 2 are both active and the remaining elements are short-circuited to ground. Identical radiation patterns occur when $\phi_{max} = 120°$, $180°$, $240°$, $300°$, and $360°$.

There have been other publications describing arrays using switched parasitic wire elements such as those described by Vaughan et al. [15, 16]. Vaughan describes a number of arrays using switched parasitic elements, from simple two-element arrays for mobile phone handsets, to more complex steerable Yagi arrays, and large circular arrays with a central active element and surrounding parasitic elements. Scott et al. [16] describe a five-element array with a single active element and four surrounding parasitic elements in which the parasitic elements can be switched between multiple loads for use in mobile communications. Other publications detailing arrays similar to those described here can be found in [17–24].

4.2.2 Microstrip Arrays

Dinger reported the first microstrip arrays using switched parasitic elements in 1982 [25–30]. Dinger also published extensions to Harrington's work on reactively controlled wire arrays [31, 32]. A 4-GHz, five-element linear microstrip array [28] with a central active patch is discussed. The remaining elements are reactively controlled parasitic patch elements. The main radiation direction ϕ_{max} is controlled by changing the reactive loads on each individual parasitic element. All elements are the same size, and the separation distance between the elements is the same. Variable phase shifters with a range from $0°$ to $250°$ requiring a bias voltage range from 0 to $-10V$ are used to adjust the reactive termination of the parasitic elements. A similar technique to that used by Harrington in [7] was used to determine the required reactive terminations. This concept was extended to a two-dimensional planar array of patch elements to achieve beam control in two directions [30].

In 1996 the concept was further modified through the use of circular microstrip elements [33]. This array consists of a central circular active element and a circular parasitic element either side of the active element. The beam is steered by controlling the reactance on each parasitic element individually.

Gray et al. [34–36] described an electronically steerable microstrip patch Yagi-Uda array. This antenna is an electronically steerable version of the mechanically steerable Yagi patch antenna developed by Huang [37, 38].

The array consists of four single Yagi microstrip patch arrays placed at 90° to each other, with a central common reflector element (see Figure 4.13). The two smaller elements in front of the active element are directors, and the larger central element is a reflector. The director and reflector elements are fixed, and the beam is steered by switching the position of the active elements using RF switches. Figure 4.13 shows a circularly polarized array where the two feed points are indicated by small circles on the active patches.

To demonstrate the electronically steerable Yagi patch array, the array presented in [34–36] has been modeled using four of the single radial arms described in [37–38]. In the array, the central reflector element is 63.5 mm square, the active elements are 55.88 mm square, and the director elements are 53.34 mm square. The separation distance between the reflector and active elements is 3.302 mm, and the distance between the active element and closest director and the two adjacent directors is 2.286 mm. Feed 1 is offset

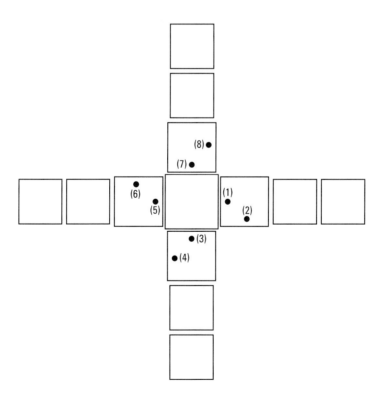

Figure 4.13 Circularly polarized, four-arm radial Yagi-Uda switched active patch antenna.

by 10.16 mm from the edge of the active element closest to the reflector, and feed 2 is offset by 15.24 mm from the side edge of the active element. The substrate is 6.35 mm thick, with $\varepsilon_r = 2.5$.

The antenna was designed for mobile satellite reception and, when modeled on Ensemble with an infinite ground plane, resulted in two resonant frequencies of 1.6123 GHz and 1.7003 GHz. Feed 1 has a 10-dB impedance bandwidth of 167 MHz (10.1%) covering both resonant frequencies. Feed 2 has a 10-dB impedance bandwidth of 73 MHz (4.5%) centered at 1.6123 GHz and 16 MHz (1%) centered at 1.7003 GHz. Figure 4.14 shows the magnitude of the reflection coefficient for both feeds in one switch position. All other switch positions will give identical results. The LHCP and RHCP radiation patterns when feed 1 and 2 are active, with 2 fed 65° out of phase from 1, are given in Figure 4.15(a) at 1.6123 GHz and in Figure 4.15(b) at 1.7003 GHz. The patterns shown are a θ cut when $\phi = 90°$. At 1.6123 GHz, the main-beam direction is $\theta_{max} = 36°$, with a maximum gain of 8.85 dBi. At 1.7003 GHz, the main-beam direction is

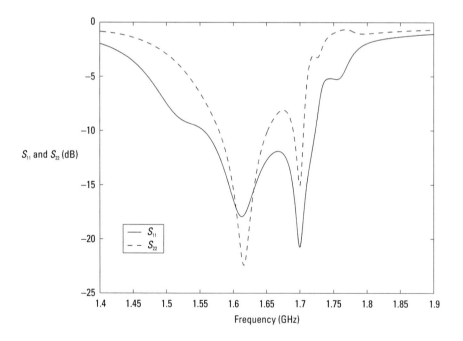

Figure 4.14 Reflection coefficient of both feeds S_{11}, S_{22} (dB) for one switch position of the circularly polarized four-arm radial Yagi-Uda switched active patch array.

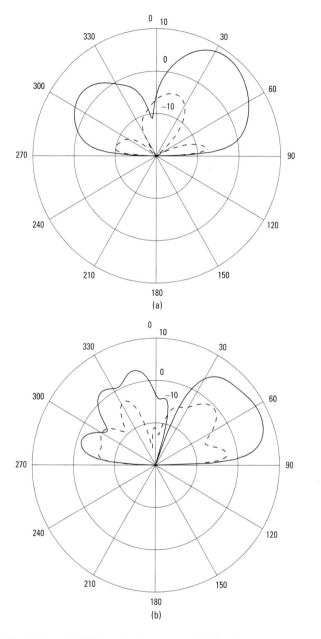

Figure 4.15 (a) LHCP and RHCP radiation pattern (dBi) of the radial arm Yagi-Uda switched active patch array at 1.6123 GHz (θ cut when $\phi = 90°$); '____' LHCP, '_ _ _' RHCP. (b) LHCP and RHCP radiation patterns (dBi) of the radial arm Yagi-Uda switched active patch array at 1.7003 GHz (θ cut with $\phi = 90°$); '____' LHCP, '_ _ _' RHCP.

$\theta_{max} = 64°$, and the maximum gain is 7.97 dBi. The front-to-back ratio for arrays with elevated beams can be calculated as: $G(\theta_{max}) - G(-\theta_{max})$. From this, the front-to-back ratio at 1.6123 GHz is 7.59 dB and at 1.7003 GHz is 10.74 dB.

The six-element circular switched wire array described by Sibille et al. [12–14] was extended to microstrip arrays [39]. A circular array of six rectangular bent-stacked slot antennas arranged in a circle form a structure similar to the monopole array [12–14]. In normal operation, two of the six elements are active and the remaining elements are terminated in a fixed $+j36\Omega$ load, to become reflectors. The main radiation direction is steered by switching the position of the active elements around the array. The paper reports a maximum directivity of 7.7 dBi. As with [12–14], this paper does not mention the possibility of using a single active element for increased beam-steering locations.

4.3   Switched Parasitic and Switched Active Wire Arrays

In Chapter 2, a two-element SASPA and a three-element FASPA were introduced. These arrays are ideal for application on cellular handsets due to their ability to direct the main beam away from the mobile user. In this section, a four-element SASPA and a five-element FASPA are introduced for use in mobile communications applications requiring greater beam-steering coverage than is achieved through the two- and three-element arrays. These arrays have been published in the literature [40–44].

4.3.1   Four-Element SASPA

A four-element monopole SASPA with dimensions $L = 0.2367\lambda_0$, $r_1 = 0.239\lambda_0$, and $t = 1.025\lambda_0$ is shown in Figure 4.16. The feed position of each element is marked with an identifier, S/Dn (n is the element number) to illustrate that each element may be a source or a parasitic element under diode control. A SASPA consists of one active element; the remaining elements are short-circuited parasitic elements. The beam is steered by switching the position of the active element using an RF switch. Table 4.4 lists the switch configurations for this array, which are required to generate four identical beams separated in azimuth by 90°.

This array was designed for operation at 1.5 GHz and constructed using wires with radius $a = 0.0059\lambda_0$. Figure 4.17 shows the measured reflection coefficient S_{11}. This shows that the impedance bandwidth of the

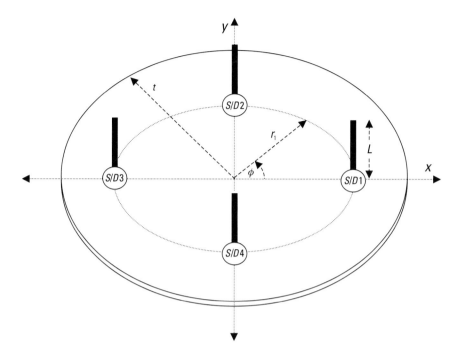

Figure 4.16 Four-element monopole SASPA.

antenna is greater than 30%. Figure 4.18 shows the measured E plane and elevated H plane radiation patterns. For antennas with an elevated beam, H plane patterns were measured by lowering the antenna under test (AUT) by the distance required, such that the angle between this antenna and the source is the beam-elevation angle. The AUT was then rotated in the same manner as a standard H plane pattern measurement. We refer to this pattern as an elevated H plane pattern. The main-beam direction is $\theta_{max} = 52°$ at all three frequencies shown, with a beamwidth $B_\phi = 90°$ at 1.35 GHz and 120° at 1.45 GHz. The front-to-back ratio for $\theta_{max} = 52°$ is 12.6 dB at 1.35 GHz and 15 dB at 1.45 GHz. The H plane radiation patterns were measured at $\theta_{max} = 52°$. The measured gain is approximately 7.5 dBi at all three frequencies. Figure 4.19 shows the measured E and H plane cross-polar radiation patterns. The cross-polarization levels are all less than -20 dBi for both the E and H planes. It is possible to obtain full 360° azimuthal coverage within 2dB of the maximum gain at 1.45 GHz or within 3 dB at 1.35 GHz (Figure 4.20).

Table 4.4
Switch Configurations of a Four-Element SASPA Required to Give Four Symmetrical Beams

Beam Direction $\phi_{max}(°)$	RF Switch State				Diode Switch State			
	$S1$	$S2$	$S3$	$S4$	$D1$	$D2$	$D3$	$D4$
0°	Closed	Open	Open	Open	Open	Closed	Closed	Closed
90°	Open	Closed	Open	Open	Closed	Open	Closed	Closed
180°	Open	Open	Closed	Open	Closed	Closed	Open	Closed
270°	Open	Open	Open	Closed	Closed	Closed	Closed	Open

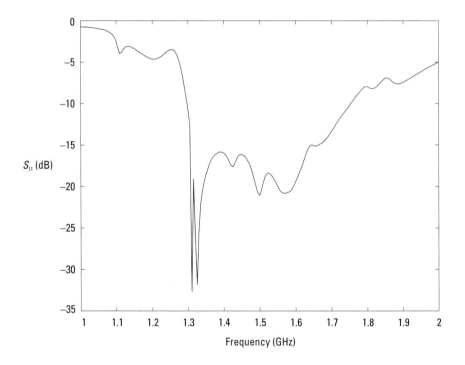

Figure 4.17 Measured frequency variation in the S_{11} (dB) of the four-element monopole SASPA shown in Figure 4.16 in one switch position.

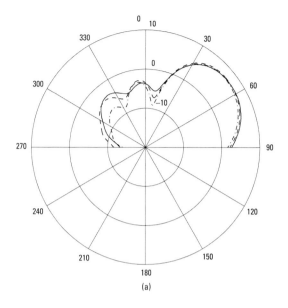

(a)

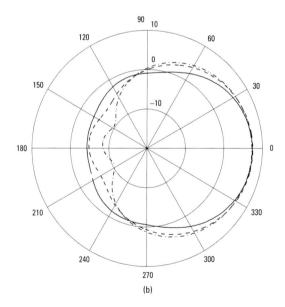

(b)

Figure 4.18 (a) Measured E plane radiation patterns (dBi) of the four-element monopole SASPA shown in Figure 4.16 in one switch position. '_____' 1.35 GHz; '_ _ _' 1.40 GHz; '_ . _' 1.45 GHz. (b) H plane radiation pattern (dBi) of the four-element SASPA shown in Figure 4.16 measured at $\theta_{max} = 52°$, for one switch position. '_____' 1.35 GHz; '_ _ _' 1.40 GHz; '_ . _' 1.45 GHz.

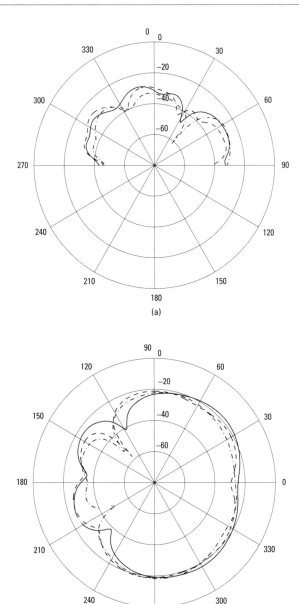

Figure 4.19 (a) Measured E plane cross-polar radiation pattern (dBi) of the four-element SASPA shown in Figure 4.16 in one switch position. '____' 1.35 GHz; '_ _ _' 1.40 GHz; '_ . _' 1.45 GHz. (b) H plane cross-polar radiation pattern (dBi) of the four-element SASPA shown in Figure 4.16 measured at θ_{max} = 52°, for one switch position. '____' 1.35 GHz; '_ _ _' 1.40 GHz; '_ . _' 1.45 GHz.

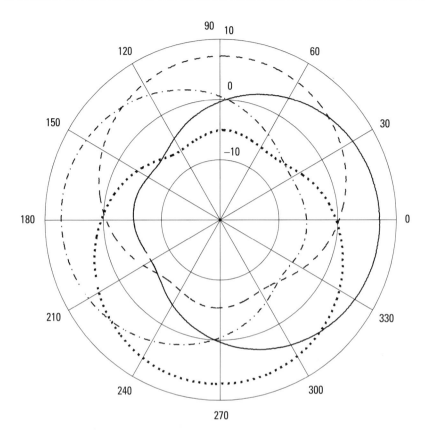

Figure 4.20 H plane radiation pattern (dBi) of the four-element SASPA shown in Figure 4.16 at 1.4 GHz measured at $\theta_{max} = 52°$, for all switch positions.

4.3.2 Five-Element FASPA

A five-element monopole FASPA on an infinite ground plane designed for 2-GHz operation is shown in Figure 4.21. The center element has length L_1 = $0.224\lambda_0$, and the parasitic elements all have the same length $L_2 = 0.2672\lambda_0$. The radius of the ring is $r = 0.25\lambda_0$. A directional beam is formed by an appropriate choice of the parasitic element termination, and the beam is steered by switching the termination on the parasitic elements around the array. The switch position of the four parasitic elements in Figure 4.21 is marked with an identifier, Dn ($n = 1,2, 3$, and 4), to illustrate that these elements are under diode control. Table 4.5 gives the switch configurations required to generate four identical beams separated by 90° in azimuth.

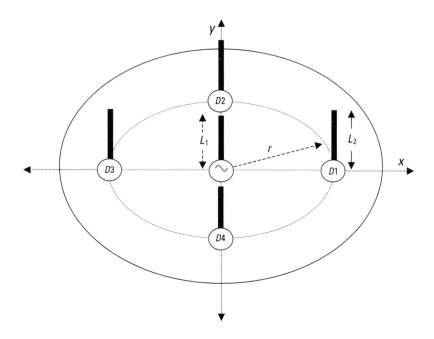

Figure 4.21 Five-element monopole FASPA on an infinite ground plane.

The array was modeled with an infinite ground plane and a wire radius of $0.0059\lambda_0$. The S_{11} is given as a function of frequency in Figure 4.22, and the H plane radiation pattern for one switch position is given in Figure 4.23. The impedance bandwidth is greater than 43%, $B_\phi = 120°$, and the front-to-back ratio is 11.7 dB. The maximum gain of the array is 9.98 dBi. The

Table 4.5
Switch Configurations of a Five-Element FASPA Required to Give Four Symmetrical Beams

Beam Direction $\phi_{max}(°)$	Diode Switch State			
	D1	D2	D3	D4
0	Open	Closed	Closed	Closed
90	Closed	Open	Closed	Closed
180	Closed	Closed	Open	Closed
270	Closed	Closed	Closed	Open

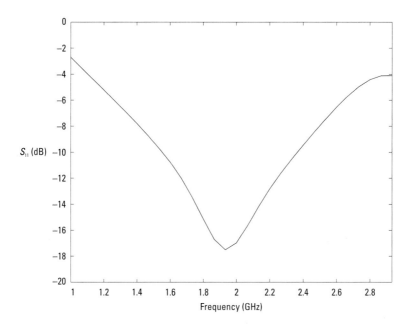

Figure 4.22 Frequency variation in S_{11} (dB) for the five-element monopole FASPA shown in Figure 4.21.

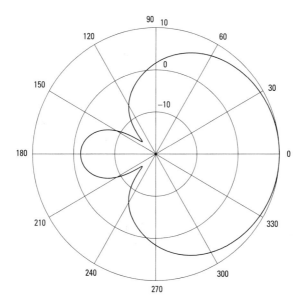

Figure 4.23 H plane radiation pattern (dBi) of the five-element FASPA shown in Figure 4.21 in one switch position at 1.9 GHz.

five-element FASPA can obtain full 360° azimuthal coverage within 1 dB of the maximum gain.

There are a number of rules that can be followed when designing an FASPA with a central active element and a single ring of switchable parasitic elements. The center frequency and the impedance bandwidth can be set by using parasitic elements slightly longer than $0.25\lambda_0$, and then reducing the length of the feed monopole from $0.25\lambda_0$. The front-to-back ratio and antenna gain are determined principally by the number of parasitic elements and the radius r of the array. Significant directionality is obtained when $r < 0.5\lambda_0$. The number of parasitic elements that are set at open circuit for a fixed beam direction should be chosen to ensure that the angle from the feed monopole is greater than 60°. Higher gain and beam control in the θ plane can be obtained if additional rings of parasitic elements are added to the array [10, 11].

4.4 Multibeam Wire Array

An array using a combination of SASPAs and FASPAs is capable of supporting multiple active elements. The array is referred to as a multibeam switched active switched parasitic array (MSASPA) [42, 44]. The array consists of a central fixed reflector element surrounded by at least two concentric rings of elements. The active element(s) lie on the inner ring, and all elements that are not active may be either open- or short-circuited parasitic elements. An example of a 13-element monopole MSASPA is given in Figure 4.24.

In order to generate a directional beam, the parasitic element located on the outer ring in line with the required direction of maximum propagation must be open circuit, and the remaining elements "behind" the active element are short-circuited. This forms a corner reflector around the active element, giving sufficient isolation for multiple simultaneous active elements. In Figure 4.24, elements 1–6 have both a diode at the feed position to switch the elements between open and short circuit and an RF switch to control the position of the active element. Elements 7–12 only have a diode at the feed position to switch between open and short circuit. Table 4.6 lists the switching configurations required to generate six single beams separated by 60° in azimuth. Table 4.7 gives the switching configurations required to generate the nine possible combinations of two simultaneous beams, and Table 4.8 gives the configurations required to generate the two combinations of three simultaneous beams.

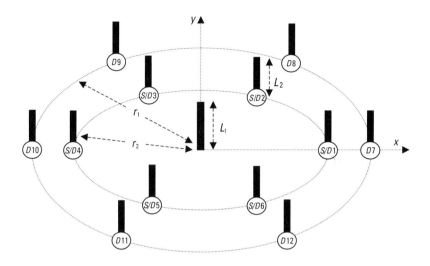

Figure 4.24 Thirteen-element monopole MSASPA.

For all MSASPAs, the number of beam-switching combinations y can be calculated from the number of simultaneous active elements x. If m is the number of switchable parasitic elements on the inner circle, the number of possible active elements is also m, and is given by:

$$m = \frac{n-1}{2} \tag{4.1}$$

where n is the total number of elements in the array. The number of unique switching solutions y for a given array is:

$$y = \frac{m}{x!} \prod_{p=1}^{x-1} (m - (2x - p)) \tag{4.2}$$

The maximum number of simultaneous active elements x_{max} is:

$$x_{max} = m / 2 \tag{4.3}$$

A 13-element monopole MSASPA with dimensions $L_1 = 0.4\lambda_0$, $L_2 = 0.2375\lambda_0$, $r_1 = 0.275\lambda_0$, and $r_2 = 0.55\lambda_0$ was constructed using a wire radius of $0.0025\lambda_0$. Figure 4.25 shows the measured reflection coefficient for a sin-

Table 4.6
Switch Configurations of a 13-Element MSASPA Required to Give Six Symmetrical Single Beams

Beam Direction $\phi_{max}(°)$		0°	60°	120°	180°	240°	300°
RF Switch State	S1	Closed	Open	Open	Open	Open	Open
	S2	Open	Closed	Open	Open	Open	Open
	S3	Open	Open	Closed	Open	Open	Open
	S4	Open	Open	Open	Closed	Open	Open
	S5	Open	Open	Open	Open	Closed	Open
	S6	Open	Open	Open	Open	Open	Closed
Diode Switch State	D1	Open	Closed	Closed	Closed	Closed	Closed
	D2	Closed	Open	Closed	Closed	Closed	Closed
	D3	Closed	Closed	Open	Closed	Closed	Closed
	D4	Closed	Closed	Closed	Open	Closed	Closed
	D5	Closed	Closed	Closed	Closed	Open	Closed
	D6	Closed	Closed	Closed	Closed	Closed	Open
	D7	Open	Closed	Closed	Closed	Closed	Closed
	D8	Closed	Open	Closed	Closed	Closed	Closed
	D9	Closed	Closed	Open	Closed	Closed	Closed
	D10	Closed	Closed	Closed	Open	Closed	Closed
	D11	Closed	Closed	Closed	Closed	Open	Closed
	D12	Closed	Closed	Closed	Closed	Closed	Open

gle active element and also for the case when there are three active elements. Figure 4.26 shows the E plane and elevated H plane radiation patterns, together with the cross-polar radiation patterns for the case when there is a single active element at three frequencies. The main-beam direction $\theta_{max} = 58°$ at all three frequencies, with a gain of 5.9 dBi at 1.36 GHz and 7.4 dBi at 1.46 GHz. The H plane radiation patterns were measured at an elevation angle of $\theta_{max} = 58°$ and the beamwidth is $B_\phi = 67°$ at 1.36 GHz and 92° at 1.46

Table 4.7
Switch Configurations of a 13-Element MSASPA Required to Give the Nine Possible Combinations of Two Simultaneous Active Elements

Beam 1 ϕ_{max} (°)		0°	0°	0°	60°	60°	60°	120°	120°	180°
Beam 2 ϕ_{max} (°)		120°	180°	240°	180°	240°	300°	240°	300°	300°
RF Switch State	$S1$	Closed	Closed	Closed	Open	Open	Open	Open	Open	Open
	$S2$	Open	Open	Open	Closed	Closed	Closed	Open	Open	Open
	$S3$	Closed	Open	Open	Open	Open	Open	Closed	Closed	Open
	$S4$	Open	Closed	Open	Closed	Open	Open	Open	Open	Closed
	$S5$	Open	Open	Closed	Open	Closed	Open	Closed	Open	Open
	$S6$	Open	Open	Open	Open	Open	Closed	Open	Closed	Closed
Diode Switch State	$D1$	Open	Open	Open	Closed	Closed	Closed	Closed	Closed	Closed
	$D2$	Closed	Closed	Closed	Open	Open	Open	Closed	Closed	Closed
	$D3$	Open	Closed	Closed	Closed	Closed	Closed	Open	Open	Closed
	$D4$	Closed	Open	Closed	Open	Closed	Closed	Closed	Closed	Open
	$D5$	Closed	Closed	Open	Closed	Open	Closed	Open	Closed	Closed
	$D6$	Closed	Closed	Closed	Closed	Closed	Open	Closed	Open	Open
	$D7$	Open	Open	Open	Closed	Closed	Closed	Closed	Closed	Closed
	$D8$	Closed	Closed	Closed	Open	Open	Open	Closed	Closed	Closed
	$D9$	Open	Closed	Closed	Closed	Closed	Closed	Open	Open	Closed
	$D10$	Closed	Open	Closed	Open	Closed	Closed	Closed	Closed	Open
	$D11$	Closed	Closed	Open	Closed	Open	Closed	Open	Closed	Closed
	$D12$	Closed	Closed	Closed	Closed	Closed	Open	Closed	Open	Open

GHz. The front-to-back ratio is 16 dB at 1.36 GHz and 12 dB at 1.46 GHz. The cross-polarization levels are all less than −20 dBi.

Figure 4.27 shows E and H plane radiation patterns for the case when there are three active elements in the array. The radiation characteristics are similar to those for the case of one active beam shown in Figure 4.25. The main-beam direction is $\theta_{max} = 60°$, with a slightly higher gain of 8 dBi. The H plane radiation pattern was measured at an elevation angle of θ_{max} and shows a beamwidth $B_\phi = 58°$ and a front-to-back ratio of greater than 12 dB for all frequencies. The cross-polarization levels are all less than −20 dBi.

Table 4.8
Switch Configurations of a 13-Element MSASPA Required to Give the Two Possible Combinations of Three Simultaneous Active Elements

Beam 1	$\phi_{max}(°)$	0°	60°
Beam 2	$\phi_{max}(°)$	120°	180°
Beam 3	$\phi_{max}(°)$	240°	300°
RF	S1	Closed	Open
Switch	S2	Open	Closed
State	S3	Closed	Open
	S4	Open	Closed
	S5	Closed	Open
	S6	Open	Closed
Diode	D1	Open	Closed
Switch	D2	Closed	Open
State	D3	Open	Closed
	D4	Closed	Open
	D5	Open	Closed
	D6	Closed	Open
	D7	Open	Closed
	D8	Closed	Open
	D9	Open	Closed
	D10	Closed	Open
	D11	Open	Closed
	D12	Closed	Open

The mutual coupling between two or more active elements in the array is of importance in duplex communications applications. When one active element is transmitting a high-power signal and the others are receiving low-power signals, there is a possibility of interference in the receiving channels. Figure 4.28 shows that the mutual coupling between the three simultaneously fed active elements is as high as -13 dB at some frequencies in the range 1.2 GHz–1.6 GHz. Across the frequency range 1.385–1.52 GHz, the coupling is always below -29 dB.

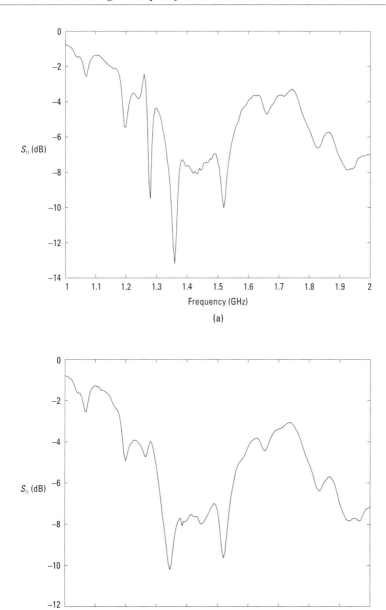

Figure 4.25 (a)S_{11} (dB) for the 13-element MSASPA shown in Figure 4.24 with a single active element. (b) S_{11} (dB) for the 13-element MSASPA shown in Figure 4.24 with three simultaneous active elements.

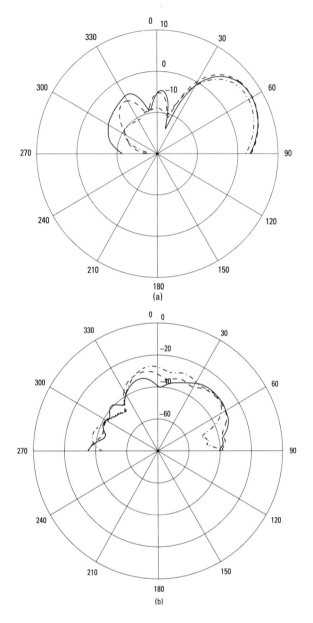

Figure 4.26 (a) Measured E plane radiation pattern (dBi) for one switch position of the 13-element MSASPA (Figure 4.24) with a single active element. '_ . _' 1.36 GHz, '_ _ _' 1.41 GHz, '____' 1.46 GHz. (b) Measured E plane cross-polar radiation pattern (dBi) for one switch position of the 13-element MSASPA (Figure 4.24) with a single active element. '_ . _ ' 1.36 GHz, '_ _ _' 1.41 GHz, '____'1.46 GHz.

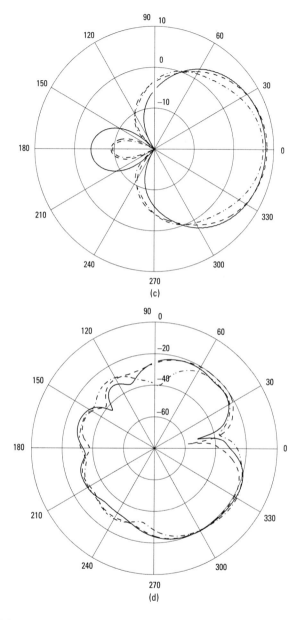

Figure 4.26 (c) H plane radiation pattern (dBi) for one switch position of the 13-element MSASPA (Figure 4.24) with a single active element measured at θ_{max}. '_ · _' 1.36 GHz, '_ _ _' 1.41 GHz, '_____' 1.46 GHz. (d) H plane cross-polar radiation pattern (dBi) for one switch position of the 13-element MSASPA (Figure 4.24) with a single active element measured at θ_{max}. '_ · _' 1.36 GHz, '_ _ _' 1.41 GHz, '_____' 1.46 GHz.

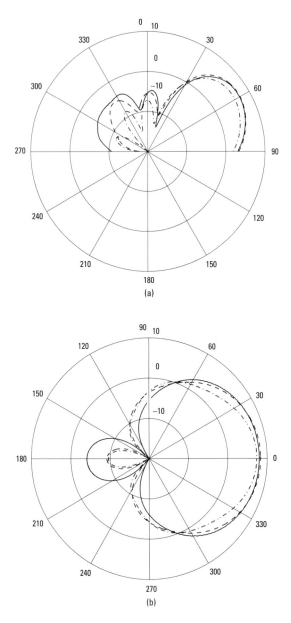

Figure 4.27 (a) Measured E plane radiation pattern (dBi) of one beam for the 13-element MSASPA (Figure 4.24) with three simultaneous active elements. '_ . _' 1.36 GHz, '_ _ _' 1.41 GHz, '_____' 1.46 GHz. (b) H plane radiation pattern (dBi) for one switch position of the 13-element MSASPA (Figure 4.24) with a single active element measured at θ_{max}. '_ . _' 1.36 GHz, '_ _ _' 1.41 GHz, '_____' 1.46 GHz.

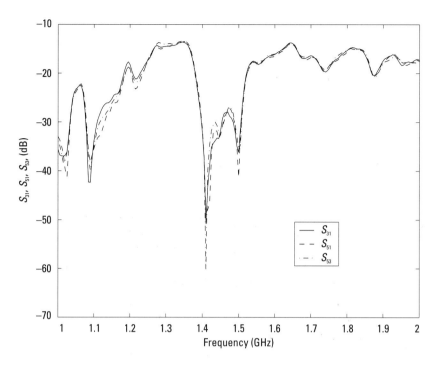

Figure 4.28 Measured mutual coupling (dB) between feed points of the 13-element MSASPA (Figure 4.24) with three simultaneous active elements.

4.5 Switched Parasitic Patch Antennas

Switched parasitic and switched active patch arrays were introduced in Chapter 3. In this section, two examples of optimized switched parasitic patch arrays are discussed. The switched parasitic patch array has been published [45]. A five-element linear array of linearly polarized patches is shown in Figure 4.29. In this array, the central active element is 41.68 mm square, the parasitic elements on either side of the active element are 41.32 mm square, and the outer parasitic elements are 40.86 mm square. The edge separation distance between the active element and the parasitic elements is 2.04 mm. The active element feed probe is centered in the x dimension and offset by 11.35 mm in the y dimension. The parasitic element short probe is also centered in the x dimension and is offset by 2.04 mm in the y dimension. The substrate is 1.56 mm thick with a dielectric constant of $\varepsilon_r = 4.35$ and a loss tangent of $\tan\delta = 0.01$.

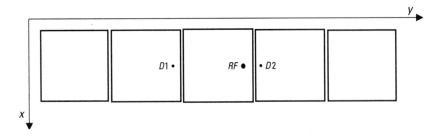

Figure 4.29 Five-element linear switched parasitic patch antenna.

In Figure 4.29, the central element feed position is marked *RF* to show that this element is always active. The parasitic elements adjacent to the active element have probes labeled *D1* and *D2* to represent the diode switching used to change the parasitic elements between short and open circuit at this point. In patch arrays, when a feed position like this is short-circuited, the element has little effect in the array, and when the feed position is open-circuited the element is a director. For this reason, it is beneficial to make the parasitic elements slightly smaller than the active element, as seen with the Yagi patch array [37, 38]. With this five-element array, there are two beam directions that can be switched, as shown in Table 4.9.

Figure 4.30 shows the measured S_{11} of the five-element switched parasitic patch array with a ground plane of 253 mm by 84 mm for both ϕ beam directions given in Table 4.9. The modeled S_{11} for an infinite ground plane is shown on the diagram. The measured results show a resonant frequency of 1.716 GHz and a 10-dB impedance bandwidth of 40 MHz (2.33%). The modeled results give a resonant frequency of 1.708 GHz and a 10-dB impedance bandwidth of 46 MHz (2.69%). The measured and modeled E plane radiation patterns at the resonant frequency of each in both switch positions are shown in Figure 4.31. The measured E plane cross-polar radiation pattern is also shown in Figure 4.31.

Table 4.9
Switch Configurations of a Five-Element Patch FASPA Required to Give Two Beams

Beam Direction	Diode Switch State	
ϕ_{max} (°)	D1	D2
90	Closed	Open
270	Open	Closed

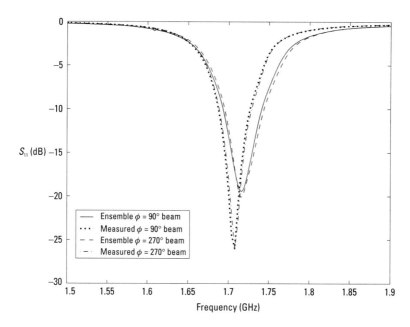

Figure 4.30 Measured and modeled S_{11} (dB) of the five-element patch array shown in Figure 4.29.

The maximum gain of the $\phi = 90°$ beam from Ensemble is 5.26 dBi at $\theta = 45°$, and the front-to-back ratio is 10.85 dB. The measured gain for the $\theta = 90°$ beam is 4.36 dBi at $\theta = 50.6°$ and the front-to-back ratio is 7.28 dB. For the $\phi = 270°$ beam from Ensemble, the maximum gain is 4.8 dBi at $\theta = -40°$ with a front-to-back ratio of 8.53 dB. The measured gain is 4.3 dBi at $\theta = -40.4°$ and the front-to-back ratio is 6.58 dB. The two switch positions do not give symmetrical radiation patterns, and the S_{11} values are also different because the array feed is not symmetrical about the center of the y axis. They are, however, similar.

Figure 4.32 shows a five-element linear circularly polarized switched parasitic array. Circular polarization is achieved through the use of two orthogonal feed probes. For this array, three shorting posts were used on each of the parasitic elements, and all three probes on each element were either open- or short-circuited simultaneously. As with the linearly polarized array, when the shorting posts are short-circuited, the element is virtually transparent, and when the posts are open-circuited, the element is a director.

This array was constructed on a 1.62-mm thick duroid substrate with $\varepsilon_r = 2.2$ and $\tan\delta = 0.0009$. The central active element is 58.26 mm square,

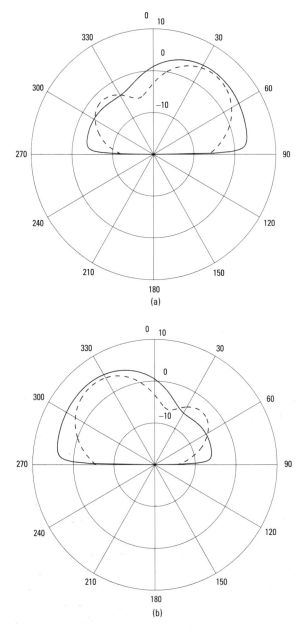

Figure 4.31 (a) E plane radiation patterns (dBi) of the patch array shown in Figure 4.29 with *D*1 closed and *D*2 open (ϕ = 90° beam). '____' Modeled, '_ _ _' Measured. (b) E plane radiation patterns (dBi) of the patch array shown in Figure 4.29 with *D*1 open and *D*2 closed (ϕ = 270° beam). '____' Modeled, '_ _ _' Measured.

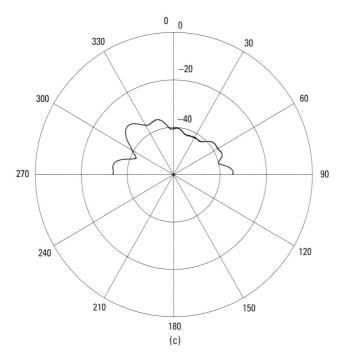

(c)

Figure 4.31 (c) Measured E plane cross-polar radiation patterns (dBi) of the patch array shown in Figure 4.29 with *D*1 closed and *D*2 open.

the parasitic elements on either side of the active element are 57.75 mm square, and the outer parasitic elements are 57.12 mm square. The edge separation distance between all adjacent elements is 2.23 mm. Feed probe *RF*1 is centered along the *x* dimension of the patch and offset by 17.61 mm from the edge of the patch in the *y* dimension. Feed probe *RF*2 is centered

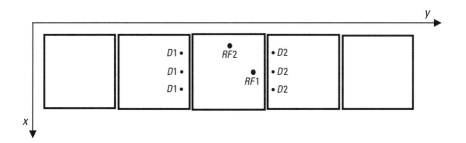

Figure 4.32 Five-element circularly polarized switched parasitic patch array.

along the y dimension and offset by 16.68 mm in the x dimension of the patch. The parasitic probes are evenly spaced along the x dimension and all are offset by 2.86 mm from the edge of the patch in the y dimension. The switch configurations given in Table 4.9 apply to this array as well.

Figure 4.33 shows the measured S_{11}, S_{22}, and S_{12} for this array in both switch positions with a ground plane size of 355.5 mm by 118.5 mm. The same scattering parameters from Ensemble with an infinite ground plane are also shown in Figure 4.33. The isolation between the feed points is good across the frequency band. The measured resonant frequency is 1.694 GHz with a 10-dB impedance bandwidth of 40 MHz (2.4 %). The resonant frequency from Ensemble is 1.706 GHz and the 10-dB impedance bandwidth is 34 MHz (2%).

The array shown in Figure 4.32 generates RHCP when feed probe $RF2$ is 90° out of phase with respect to feed probe $RF1$. The RHCP radiation patterns have been measured at the resonant frequency in both switch positions using the rotating-source method [46]. Figure 4.34 shows the measured RHCP radiation patterns in each switch position. The Ensemble patterns at the resonant frequency are also shown.

The maximum gain of the $\phi_{max} = 90°$ beam from Ensemble is 8.955 dBi at $\theta = 28°$ and the front-to-back ratio of this beam is 11.92 dB. The measured gain for this beam is 9.16 dBi at $\theta = 29°$ with a front-to-back ratio of 13.05 dB. For the $\phi_{max} = 270°$ beam, the maximum gain from Ensemble is 8.984 dBi at $\theta = -29°$ with a front-to-back ratio of 12.37 dB. The measured gain is 9.56 dBi at $\theta = -26.2°$ with a front-to-back ratio of 12.79 dB.

4.6   Dielectric Coated and Dielectric Resonator Antennas

The size of a resonant wire antenna can be reduced by embedding the conducting elements in a dielectric or magnetic material with a relative permittivity and/or relative magnetic permeability greater than unity [46–48]. A perfectly conducting thin wire antenna embedded in a lossless material of infinite extent has a fundamental resonance when it is a half-wave dipole; that is, the length $l = \lambda_m / 2$ where λ_m is the wavelength in the lossless material given by:

$$\lambda_m = \frac{c}{f\sqrt{\varepsilon_r \mu_r}} \qquad (4.4)$$

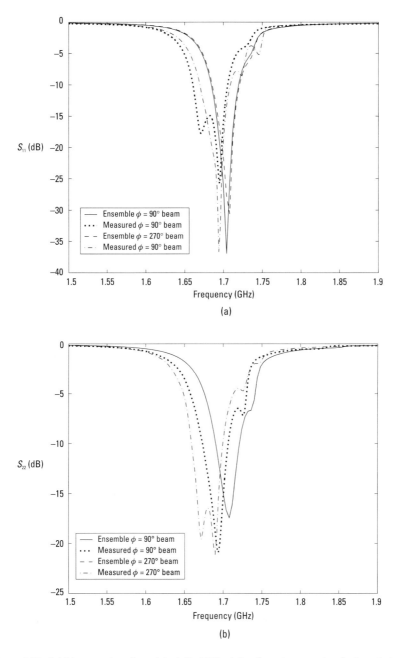

Figure 4.33 (a) Measured and modeled S_{11} (dB) of the five-element circularly polarized FASPA shown in Figure 4.32. (b) Measured and modeled S_{22} (dB) of the five-element circularly polarized FASPA shown in Figure 4.32.

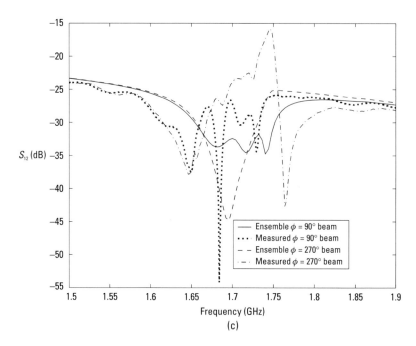

Figure 4.33 (c) Measured and modeled S_{12} (dB) of the five-element circularly polarized FASPA shown in Figure 4.32.

where c is the velocity of light in a vacuum, f is the frequency of the radiation, ε_r is the relative permittivity, and μ_r is the relative permeability of the medium. In free space $\varepsilon_r = 1$ and $\mu_r = 1$, and so λ_m reduces to λ_0.

The half-wave resonance condition occurs when the length of the wire l is given by:

$$l = \frac{\lambda_0}{2\sqrt{\varepsilon_r \mu_r}} \qquad (4.5)$$

In most communications situations, the region between the transmitting and receiving antennas is free space (i.e., air) and so the embedding material surrounding one antenna is not of infinite extent. It is still possible to reduce the length of the wire element by coating it with a layer of dielectric or magnetic material that has a finite width. In this case, (4.4) is no longer accurate and the resonant length of the antenna depends on the thickness of the coating as well as ε_r and μ_r.

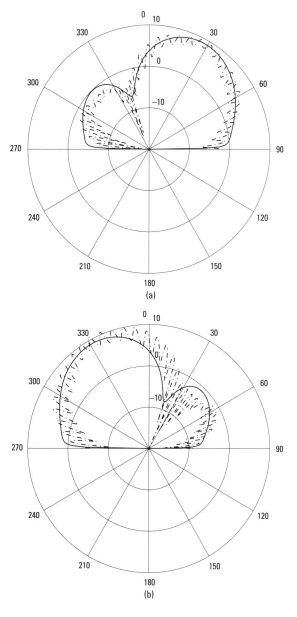

Figure 4.34 (a) Measured and modeled RHCP radiation patterns (dBi) of the circularly polarized FASPA shown in Figure 4.32 with $D1$ closed and $D2$ open (θ cut when $\phi = 90°$). '____' Modeled, '_ _ _' Measured. (b) Measured and modeled RHCP radiation patterns (dBi) of the circularly polarized FASPA shown in Figure 4.32 with $D1$ open and $D2$ closed (θ cut when $\phi = 270°$). '____' Modeled, '_ _ _' Measured.

When the coating is thin compared to $0.5\lambda_m$, only one surface wave mode can propagate, and the effective length of the antenna is independent of the type of feed. In this case, the antenna can be treated in a manner similar to wire antenna structures in free space, but with the effective length and separation between the wires larger than the true dimensions.

When the coating is greater than or equal to $0.5\lambda_m$, more than one surface-wave mode and a number of cavity modes are possible. The location of the feed point is significant in determining which modes are launched. The radiation characteristics are dominated by resonances in the dielectric material alone rather than in the wire elements. This class of antennas is referred to as dielectric resonator antennas (DRA). These two antenna technologies are discussed separately.

4.6.1 Dielectric Coated Wire Antennas

Consider the case of a monopole antenna with radius a, on a ground plane of infinite extent (Figure 4.35). The monopole is located at the center of a dielectric cylinder having outer radius b and electromagnetic properties ε_r and

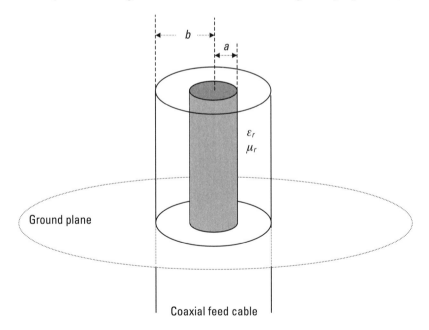

Figure 4.35 Embedded monopole antenna on an infinite ground plane.

μ_r. The length of the cylinder is identical to that of the monopole. The monopole is connected to the center conductor of a coaxial feed cable located beneath the ground plane, and the outer conductor is connected directly to the ground plane.

A thin material coating supports a TM surface wave [46–48]. This is equivalent to the surface wave mode that propagates in a dielectric slab on top of a conducting plane [48]. This propagation mode is a slow wave mode, that is, it has a velocity less than the velocity of light, and so the effective length of the structure is longer than the free space equivalent. In the radial direction, the field strength decreases exponentially away from the surface of the material. At the end of the monopole, the surface wave is mainly reflected, although there is radiation from the end. The phase constant of the surface wave mode β can be solved from a set of transcendental equations involving Bessel functions [48]:

$$\frac{K_1(pb)}{pK_0(pb)} = \frac{\varepsilon_r}{h} \frac{J_0(ha)Y_1(hb) - J_1(hb)Y_0(ha)}{J_0(hb)Y_0(ha) - J_0(ha)Y_0(hb)} \tag{4.6}$$

where

K_0 and K_1 are modified Bessel functions of the second kind;

J_0 and J_1 are Bessel functions of the first kind;

Y_0 and Y_1 are Bessel functions of the second kind;

ε_r is the relative permittivity of the coating material;

the wave numbers $k_0 = 2\pi/\lambda_0$, $k = 2\pi/\lambda_m$;

and

$$p^2 = \beta^2 - k_0^2 \quad \text{and} \quad h^2 = k^2 - \beta^2 \tag{4.7}$$

As the thickness increases, $\lambda_m \to \lambda_0 / \sqrt{\varepsilon_r}$. This limiting case is that for a wire embedded in a material of infinite extent given by (4.4).

Figure 4.36 illustrates the variation in the effective wavelength along the coated wire as a function of both dielectric thickness and relative permittivity. The central wire has a radius of 1 mm, and the conductivity of the wire is assumed to be infinite. The frequency is 2 GHz.

When a conductor lies on a plane interface between two lossless dielectric materials of infinite extent, the effective relative permittivity ε_{eff} can

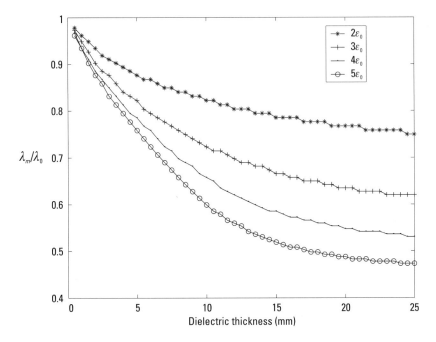

Figure 4.36 Variation in the wavelength λ_m relative to the free space wavelength λ_0, for a dielectric coating of variable thickness (b-a) and relative permittivity ε_r. The frequency is 2 GHz and the wire radius is 1 mm.

be approximated as the average of the two relative permittivity values ε_{r1} and ε_{r2} [49]:

$$\varepsilon_{\text{eff}} = \frac{\varepsilon_{r1} + \varepsilon_{r2}}{2} \tag{4.8}$$

When the coating is not radially symmetric about the wire conductor, the H plane radiation pattern is directional [50] even when there are no resonances present in the dielectric or ferrite material. This method of reducing the size of dipole elements has been applied to microstrip dipoles covered with a dielectric slab [51].

A dielectric cylinder SASPA was designed using (4.5) and (4.6) with four elements lying on the surface of a dielectric cylinder [52]. The antenna uses a single switched active element with the other three elements acting as reflectors. The three parasitic elements form a rudimentary corner reflector. This is similar to a four-element SASPA in free space. A monopole structure is illustrated in Figure 4.37, with element length l and cylinder diameter D. The relative permittivity of the dielectric material was $\varepsilon_r = 3.3$. An antenna with $l = 76$ mm and $D = 25$ mm was designed for operation at 700 MHz. The array has been modeled on HFSS [53] as a dipole array, and the H plane radiation pattern at 700 MHz is given in Figure 4.38 in one switch position. All other switch positions give identical beams. The antenna has four switch positions and each has a front-to-back ratio of 8.7 dB and an approximate beamwidth $B_\phi = 130°$.

A nine-element monopole SASPA antenna was designed to operate at 2.4 GHz [54]. The array was manually optimized for front-to-back ratio and input impedance using a finite-difference time-domain model. The dielectric material was L6G nylon, which has a relative permittivity of $\varepsilon_r = 4.4$. The nylon cylinder has a diameter of $0.65\lambda_m$ and a length of $0.33\lambda_m$. The radial element spacing of the two circles of parasitic elements from the central

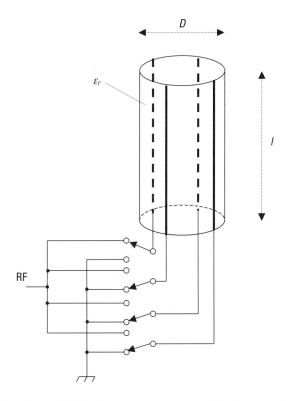

Figure 4.37 Four-element SASPA for a cellular telephone handset from Thiel et al. [52]. The electronic control lines are not drawn to scale.

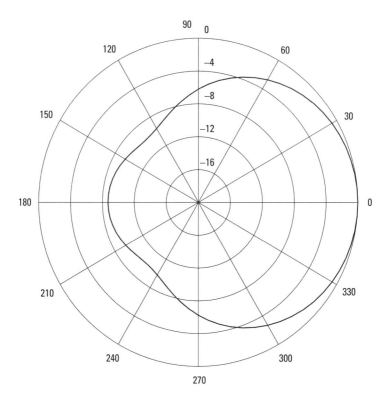

Figure 4.38 H plane radiation pattern (normalized dB) of a dielectric dipole SASPA modeled on HFSS.

reflector element is $r_1 = 0.2\lambda_m$ and $r_2 = 0.4\lambda_m$. The radial elements are $0.25\lambda_m$ long and the center-element length is the same as the length of the cylinder. The structure is shown schematically in Figure 4.39. The measured reflection coefficient and the H plane radiation pattern from the finite-difference time-domain model are given in Figures 4.40 and 4.41, respectively. The antenna has a front-to-back ratio of 21 dB and a beamwidth of $B_\phi = 180°$ [54].

The size of this antenna is smaller than a nine-element SASPA monopole array in air that has a similar front-to-back ratio and impedance bandwidth [54]. Until the radiation pattern measurements are performed, any degradation in the gain for the array embedded in dielectric will not be fully known. The presence of the dielectric material has reduced the area of the antenna footprint by 32%. The dielectric material also offers the structure mechanical rigidity, although the total weight of the antenna has been increased.

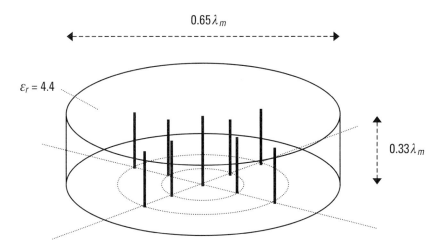

Figure 4.39 SASPA embedded in nylon for 2.4-GHz operation from [54].

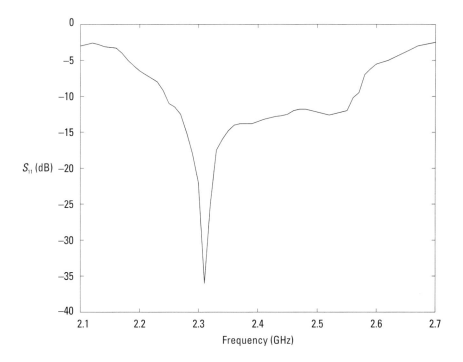

Figure 4.40 Measured S_{11} (dB) for the SASPA embedded in nylon (Figure 4.39).

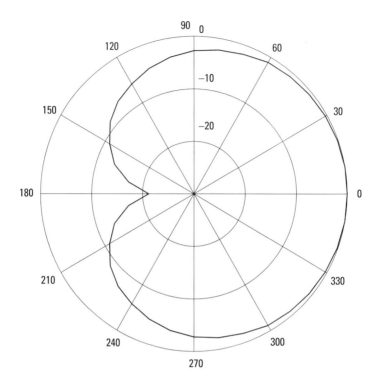

Figure 4.41 Modeled H plane radiation pattern (normalized dB) for the SASPA embedded in nylon (Figure 4.39) at 2.4 GHz using a finite-difference time-domain
model.

4.6.2   Dielectric Resonator Antennas

At high frequencies, the size of antenna structures becomes very small and
the appropriate scaling of the conducting elements makes construction difficult. The length, diameter, and spacing of wire elements are critical and subject to fine mechanical tolerance. The depth of penetration of the current
along the wire (the skin depth) is related to the frequency [48]. This is discussed in Chapter 6. The net result is that the wire resistance increases as the
frequency increases. The finite conductivity of wire elements of small diameter exacerbates this effect.

To overcome these problems, it was suggested that a resonant dielectric
cavity could serve as a radiating element [55]. The antenna consists of a
dielectric volume lying on a ground plane fed by a short conducting probe
located inside the dielectric or an open-circuit microstip line located immediately below the dielectric. The shape of the DRA can be cylindrical [55],

hemispherical [56], or rectangular [57]. This DRA is similar in principle to a metal cavity resonator where there is field leakage through the dielectric-air interface. The radiation pattern and the input impedance of the DRAs can be calculated analytically using the boundary conditions appropriate to the resonant cavity mode excited [55–57]. A cylindrical DRA fed with a feed probe offset from the central position has a figure-of-eight H plane polar pattern [55] when excited at the fundamental $HEM_{11\delta}$ mode.

It is possible to switch the main-beam direction in the H plane by changing the feed position to one of a number of probes already located inside the dielectric [58] (see Figure 4.42). The unused probe feeds are not connected to the ground plane and so play little role in the overall characteristics of the antenna similar to the open-circuit elements in FASPAs and SASPAs. Kingsley and O'Keefe [59] extended this concept by using a power splitter/power combiner to feed more than one probe simultaneously. This antenna was designed for monopulse radar applications at VHF. In this way, the number of beam positions can be increased significantly, and the array has characteristics similar to that of a phased array. When the power splitter provides unequal power to the two probes, additional positions of the main beam can be achieved [59]. As the antenna beamwidth is quite large, the shift in the position of the nulls in the radiation pattern is more important for maximizing the signal-to-noise ratio rather than a significant change in gain at angles close to the main-beam direction.

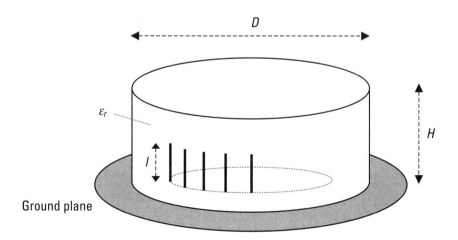

Figure 4.42 Electronically steerable DRA.

A switchable DRA structure was designed for operation at approximately 55 MHz using water as the dielectric material. The water was contained in a PVC cylinder (diameter $D = 550$ mm and height $H = 200$ mm) mounted on a small, conductive ground plane. The length of the probes was approximately $l = 100$ mm. The 10-dB impedance bandwidth was measured to be approximately 1 MHz [59]. As the relative permittivity of water is $\varepsilon_r = 84$ at a temperature of 15°C, a significant decrease in antenna size was achieved.

4.7   Tin-Can Antenna

It is often desirable to incorporate antennas into existing infrastructure or to design antennas that are unobtrusive in the urban environment. The advantages in this approach include improved aesthetics, a reduction in the number of attacks by vandals, reduced wind resistance, and reduced costs associated with the installation of additional support structures.

The tin-can antenna is a five-element monopole FASPA antenna designed to fit inside a 10-cm-diameter PVC pole (see Figure 4.43). Problems associated with the finite ground plane (see Section 2.6) are reduced by the addition of a conducting cylinder with a length of approximately $\lambda_0/4$. The design frequency was 1.36 GHz and so the monopoles are 55 mm long with a radius of 2 mm and are mounted on a flat circular ground plane of radius 32 mm, that forms the top of a hollow cylinder with length $t = 70$ mm. The control electronics can be housed immediately beneath the ground plane. The ground plane therefore resembles a tin can with the feed cable end opened. The four parasitic elements lie very close to the edge of the ground plane, and the length of the cylinder is optimized to control the elevation angle of the radiation pattern. The resonant frequency was measured to be $f_0 = 1.39$ GHz. The tin-can antenna is an extension of a coaxially fed monopole with a conducting sleeve shown in Figure 2.15(a). The cylinder provides a quarter-wave resonant ground reference, giving a performance similar to an infinite ground plane, but with the additional benefit of being able to control the elevation angle of the radiation.

In many situations it is desirable that the antenna is located above the level of the surrounding objects so that line-of-sight illumination of the area is possible. For many applications, the beam direction should lie close to or even beneath the horizontal plane.

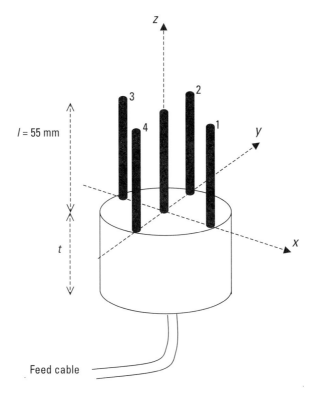

Figure 4.43 Center-fed FASPA tin-can antenna.

In Figure 2.17 it is evident that the effect of a finite circular ground plane on a single monopole antenna is to elevate the radiation above the horizontal plane. The radiation can be directed toward the horizontal plane by increasing the length of the tin-can t. This has only a minor effect on the resonant frequency of the antenna.

In Figure 4.43, if element 1 is open circuit and elements 2, 3, and 4 are short circuit, the radiation is directed in the $\phi = 0°$ direction. In this case $\theta_{max} = 90°$ and $B_\theta = 90°$. Figure 4.44 shows the measured E plane radiation pattern in the plane of maximum radiation for the tin-can antenna with the measured results for a similar array without the vertical ground plane extension. Figure 4.45 shows the H plane radiation pattern of the tin-can antenna in the horizontal plane (for $\theta = 90°$). The tin-can antenna has a front-to-back ratio of better than 12 dB and achieves the desired goal of directing the radiation into the horizontal plane.

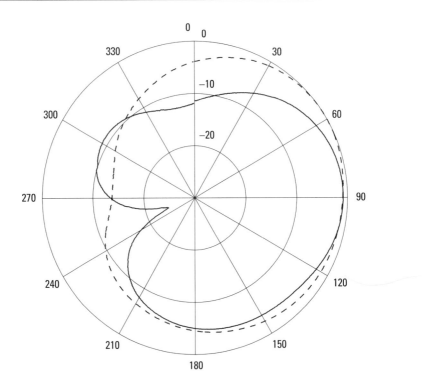

Figure 4.44 Measured E plane radiation patterns (normalized dB) for a center-fed FASPA antenna with the tin can and with a small circular ground plane. '____' tin can; '_ _ _' ground plane.

4.8 Parabolic-Antenna Beam Steering

Parabolic dish antennas are used for large-aperture, narrow-beamwidth point-to-point microwave links. A metallic parabolic reflector with a feed antenna located at its focus can have a beamwidth of less than 1° [60]. The basic structure of a single reflector parabolic dish antenna is shown in Figure 4.46. The line-of-sight path between two microwave antennas is affected by the temperature and the humidity of the intervening atmosphere [61]. This clear air diffraction leads to fading. It is sometimes advantageous to change the direction of the main beam into the position of the first null in the radiation pattern. As the atmosphere is horizontally layered, the requirement is for a beam shift in the vertical plane. While it is possible to move the complete antenna structure mechanically, this requires significant power. It

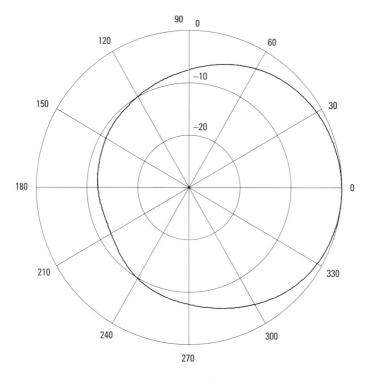

Figure 4.45 H plane radiation pattern (normalized dB) of the tin-can antenna measured at 1.39 GHz.

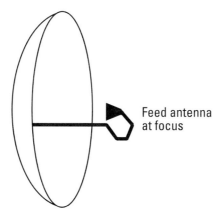

Figure 4.46 Parabolic-dish reflector antenna used in line-of-sight microwave links. The feed is located at the focus.

also makes precise positioning of the antenna difficult in conditions of high wind.

Ruze [62] demonstrated that the principal beam direction is changed slightly by mechanically moving the feed position in the vertical plane. Such a movement, however, substantially degrades the performance of the antenna in terms of beamwidth and gain, as the feed is no longer at the focus. An alternate solution proposed by Durnan [63] uses a switched parasitic antenna located at the focus as the feed. Figure 4.47 illustrates a parabolic dish antenna with two parasitic elements displaced vertically and symmetrically about the driven dipole feed element. The driven antenna remains at the focal point of the dish. When the top parasitic element is short-circuited, significant current is induced in this element. The total radiation pattern is the summation of the two components. The beam can be adjusted by switching the parasitic elements to increase the radiated power in the position of the first null. Figure 4.48 shows the measured H plane radiation patterns of the array shown in Figure 4.47, with a 32-cm dish for the case when both parasitic dipoles are open-circuited and also when the lower parasitic dipole is short-circuited. This shows that the beam is shifted by approximately 10° in the ϕ plane with little or no degradation in the gain and beamwidth. These patterns were measured at 1 GHz.

The optimal position of a vertically displaced parasitic element was found to be at approximately $0.5\lambda_0$. With this separation distance, the current magnitude in the parasitic element is approximately 40% of that in the feed antenna, with a phase delay of slightly less than 90°. The effect of the

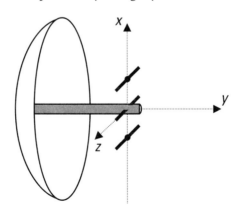

Figure 4.47 Switched parasitic feed for a parabolic dish antenna. The two parasitic elements are displaced vertically from the RF-driven element and have a diode switch located at their center. The vertical polar pattern is controlled electronically by changing the impedance state of the diodes [63].

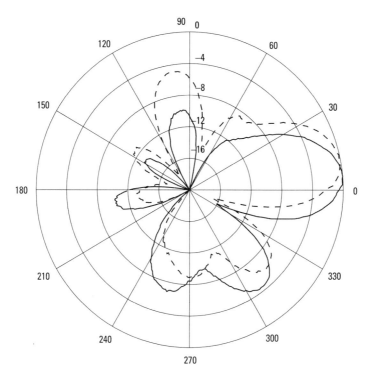

Figure 4.48 Measured H plane radiation patterns (normalized dB) of a 32-cm dish with a switched parasitic feed. '____' both parasitic elements open circuit; '_ _ _' lower parasitic short-circuited.

parasitic elements on the cross-polar radiation is minimal and the gain of the antenna increases slightly.

The current induced in the shorted parasitic element is phase shifted with respect to the driven element and so the performance of the antenna can be further improved through both a vertical and horizontal optimization of the position. In order to preserve the impedance matching of the feed antenna, the parasitic elements must be symmetrically placed around the feed. When the parasitic element is closer to the dish than the feed element, the phase path to the antenna aperture is reduced and the phase delay in the current in the parasitic element decreases. The compromise is that the parasitic element can be located closer to the feed element with consequently higher current [64].

This structure is mechanically robust and electronically controllable. It can be retrofitted to existing parabolic dish antennas with no significant change to the RF circuit. The beam can be controlled by digital logic voltages from a microcontroller circuit.

4.9 Summary

A variety of switched parasitic antennas have been described in this chapter. The technique has applications to cellular telephone handsets, cellular base stations, satellite communications, radio navigation, and microwave links. In most cases, the total size of the antennas is less than one wavelength and the radiation patterns have beamwidths greater than 90°. Smaller beamwidths can be achieved by using larger arrays of parasitic elements with the requirement that the distance between adjacent elements must be within approximately $\lambda_0/2$ for significant mutual coupling. While the range in beamwidth is relatively small, the range in bandwidth is large, and varies from quite narrow to greater than 10%. This is influenced by the variation in element size within the array. There are similarities between these switched parasitic antennas and Yagi-Uda antennas.

The design procedure for switched parasitic antennas is a complicated process requiring the optimization of a large number of parameters. Chapter 5 presents a number of techniques that can be used to achieve the best antenna performance for a specific application.

References

[1] NEC WIN Professional TM, Antenna Analysis Software version 1.1, Nittany Scientific Inc., California, 1997.

[2] Ansoft Corporation, Ensemble, 2000, http://www.ansoft.com.

[3] Himmel, L., S. H. Dodington, and E. G. Parker, Electronically Controlled Antenna System, U.S. Patent No. 3560978, February 2, 1971.

[4] Black, S. H., and R. B. Formeister, Direction Finder System, U.S. Patent No. 3725938, April 3, 1973.

[5] Gueguen, M., Electronically Step-by-Step Rotated Directive Radiation Beam Antenna, U.S. Patent No. 3846799, November 5, 1974.

[6] Harrington, R. F., "Reactively Controlled Antenna Arrays," *IEEE APS International Symposium Digest*, Amherst, MA, 1976, pp. 62–65.

[7] Harrington, R. F., "Reactively Controlled Directive Arrays," *IEEE Trans. Antennas and Propagation*, Vol. 26, No. 3, 1978, pp. 390–395.

[8] Luzwick, J., and R. F. Harrington, "Reactively Loaded Aperture Antenna Array," *IEEE Trans. Antennas and Propagation*, Vol. 26, No. 4, 1978, pp. 543–547.

[9] Dumas, T. A., and L. V. Griffe, Electronically Rotated Antenna Apparatus, U.S. Patent No. 4631546, December 23, 1986.

[10] Milne, R. M. T., "A Small Adaptive Array Antenna for Mobile Communications," *IEEE APS International Symposium Digest*, 1985, pp. 797–800.

[11] Milne, R., Adaptive Array Antenna, U.S. Patent No. 4700197, October 13, 1987.

[12] Sibille, A., C. Roblin, and G. Poncelet, "Circular Switched Monopole Arrays for Beam Steering Wireless Communications," *Electronics Letters*, Vol. 33, No. 7, 1997, pp. 551–552.

[13] Sibille, A., et al., "Beam Steering Circular Arrays for WLANs at 5 and 17 GHz," *NICE—International Symposium Antennas*, 1998, pp. 354–357.

[14] Chelouah, A., et al., "Angular Diversity Based on Beam Switching of Circular Arrays for HIPERLAN Terminals," *Electronics Letters*, Vol. 36, No. 5, 2000, pp. 387–388.

[15] Vaughan, R., "Switched Parasitic Elements for Antenna Diversity," *IEEE Trans. Antennas and Propagation*, Vol. 47, No. 2, 1999, pp. 399–405.

[16] Scott, N. L., M. O. Leonard-Taylor, and R. G. Vaughan, "Diversity Gain from a Single-Port Adaptive Antenna Using Switched Parasitic Elements Illustrated with a Wire and Monopole Prototype," *IEEE Trans. Antennas and Propagation*, Vol. 47, No. 6, 1999, pp. 1066–1070.

[17] Chisholm, J. P., Stepped Cardioid Bearing System, U.S. Patent No. 3950753, April 13, 1976.

[18] Parker, E. G., Antenna Pattern Synthesis and Shaping, U.S. Patent No. 4260994, April 7, 1981.

[19] Henderson, A. S., Antenna Having Electrically Positionable Phase Center, U.S. Patent No. 4387378, June 7, 1983.

[20] Audren, J., and P. Brault, High Frequency Antenna with a Variable Directing Radiation Pattern, U.S. Patent No. 5235343, August 10, 1993.

[21] Sanford, G. G., and P. M. Westfeldt, Electronically Reconfigurable Antenna, U.S. Patent No. 5294939, March 15, 1994.

[22] Pritchett, D. M., Communication System and Methods Utilizing a Reactively Controlled Directive Array, U.S. Patent No. 5767807, June 16, 1998.

[23] Taenzer, J. C., Adjustable Array Antenna, U.S. Patent No. 5905473, May 18, 1999.

[24] Koscica, T. E., and B. J. Liban, Azimuth Steerable Antenna, U.S. Patent No. 6037905, March 14, 2000.

[25] Dinger, R. J., "Adaptive Microstrip Antenna Array Using Reactively Terminated Parasitic Elements," *IEEE APS International Symposium Digest*, Albuquerque, NM, 1982, pp. 300–303.

[26] Dinger, R. J., "A Microstrip Power Inversion Array Using Parasitic Elements," *IEEE APS International Symposium Digest*, Houston, TX, 1983, pp. 191–194.

[27] Dinger, R. J., "A Computer Study of Interference Nulling by Reactively Steered Adaptive Arrays," *IEEE APS International Symposium Digest*, Boston, MA, 1984, pp. 807–810.

[28] Dinger, R. J., "Reactively Steered Adaptive Array Using Microstrip Patch Elements at 4 GHz," *IEEE Trans. Antennas and Propagation*, Vol. 32, No. 8, 1984, pp. 848–856.

[29] Dinger, R. J., "A Planar 4.0 GHz Reactively Steered Adaptive Array," *IEEE MTT-S International Symposium Digest*, San Francisco, CA, 1984, pp. 303–305.

[30] Dinger, R. J., "A Planar Version of a 4.0 GHz Reactively Steered Adaptive Array," *IEEE Trans. Antennas and Propagation*, Vol. 34, No. 3, 1986, pp. 427–431.

[31] Dinger, R. J., and W. D. Meyers, "Compact Reactively Steered Antenna Array," *IEEE APS International Symposium Digest*, Quebec, Canada, 1980, pp. 312–315.

[32] Dinger, R. J., "Simulation Study of Reactively Steered Adaptive Arrays," *Electronics Letters*, Vol. 21, No. 9, 1985, pp. 383–384.

[33] Cailleu, D., N. Haese, and P. A. Rolland, "Microstrip Adaptive Array Antenna," *Electronics Letters*, Vol. 32, No. 14, 1996, pp. 1246–1247.

[34] Gray, D., J. W. Lu, and D. V. Thiel, "Electronically Steerable Yagi-Uda Microstrip Patch Antenna Array," *IEEE APS International Symposium Digest*, Newport Beach, CA, 1995, p. 1870.

[35] Gray, D. P., J. W. Lu, and L. Shafai, "Experimental Study of Parasitically Steered Fixed Beam Microstrip Patch Arrays," *IEEE APS International Symposium Digest*, Montreal, Canada, 1997, pp. 1276–1279.

[36] Gray, D., J. W. Lu, and D. V. Thiel, "Electronically Steerable Yagi-Uda Microstrip Patch Antenna Array," *IEEE Trans. Antennas and Propagation*, Vol. 46, No. 5, 1998, pp. 605–608.

[37] Huang, J., "Planar Microstrip Yagi Array Antenna," *IEEE APS International Symposium Digest*, San José, CA, 1989, pp. 894–897.

[38] Huang, J., and A. C. Densmore, "Microstrip Yagi Array Antenna for Mobile Satellite Vehicle Applications," *IEEE Trans. Antennas and Propagation*, Vol. 39, No. 7, 1991, pp. 1024–1030.

[39] Fassetta, S., and A. Sibille, "Switched Angular Diversity BSSA Array Antenna for WLAN," *Electronics Letters*, Vol. 36, No. 8, 2000, pp. 702–703.

[40] Preston, S. L., and D. V. Thiel, "Direction Finding Using a Switched Parasitic Antenna Array," *IEEE APS International Symposium Digest*, Montreal, Canada, 1997, pp. 1024–1027.

[41] Preston, S. L., et al., "Base-Station Tracking in Mobile Communications Using a Switched Parasitic Antenna Array," *IEEE Trans. Antennas and Propagation*, Vol. 46, No. 6, 1998, pp. 841–844.

[42] Preston, S. L., J. W. Lu, and D. V. Thiel, "Systematic Approach to the Design of Directional Antennas Using Switched Parasitic and Switched Active Elements," *Asia Pacific Microwave Conference Proceedings*, Yokohama, Japan, 1998, pp. 531–534.

[43] Preston, S. L., and D. V. Thiel, "Size Reduction of Switched Parasitic Directional Antennas Using Genetic Algorithm Optimisation Techniques," *Asia Pacific Microwave Conference Proceedings*, Yokohama, Japan, 1998, pp. 1401–1404.

[44] Preston, S. L., D. V. Thiel, and J. W. Lu, "A Multibeam Antenna Using Switched Parasitic and Switched Active Elements for Space-Division Multiple Access Applications," *IEICE Trans. Electronics*, Vol. E82-C, No. 7, 1999, pp. 1202–1210.

[45] Preston, S. L., et al., "Electronics Beam Steering Using Switched Parasitic Patch Elements," *Electronics Letters*, Vol. 33, No. 1, 1997, pp. 7–8.

[46] James, J. R., and A. Henderson, "Electrically Short Monopole Antennas with Dielectric or Ferrite Coatings." *Proc. IEE,* Vol. 125, September 1978, pp. 793–803.

[47] James, J. R., A. J. Schuler, and R. F. Binham, "Reduction of Antenna with Dielectric Loading," *Electronics Letters*, Vol. 10, No. 2, 1974, pp. 263–265.

[48] Collin, R. E., *Field Theory of Guided Waves*, 2nd ed., New York: IEEE Press, 1991.

[49] Knight, P., "Low-Frequency Behaviour of the Beverage Aerial," *Electronics Letters,* Vol. 13, No. 1, 1977, pp. 21–22.

[50] King, R. W. P., "The Many Faces of the Insulated Antenna," *Proc. IEEE*, Vol. 64, 1976, pp. 228–238.

[51] Kishioka, N., and H. Arai, "FDTD Analysis of Strip Dipole Antenna Covered by Dielectric Material," *Asia Pacific Microwave Conference Proceedings*, Sydney, Australia, December 2000, pp. 1352–1355.

[52] Thiel, D. V., S. G. O'Keefe, and J. W. Lu, Antennas for Use in Portable Communications Devices, U.S. Patent No. 6034638, March 2000.

[53] Ansoft Corporation, HFSS, 2000, http://www.ansoft.com.

[54] Lu, J., et al., "Multibeam Switched Parasitic Antenna Embedded in Dielectric for Wireless, Communications Systems," *Electronics Letters*, Vol. 37, No. 14, 2001, pp. 871–872.

[55] Long, S. A., M. W. McAllister, and L. C. Shen, "The Resonant Cylindrical Dielectric Cavity Antenna," *IEEE Trans. Antennas and Propagation,* Vol. 31, No. 3, May 1983, pp. 406–412.

[56] McAllister, M. W., and S. A. Long, "Resonant Hemispherical Dielectric Antenna," *Electronics Letters,* Vol. 20, No. 16, August 1984, pp. 657–659.

[57] McAllister, M. W., S. A. Long, and G. L. Conway, "Rectangular Dielectric Antenna," *Electronics Letters,* Vol. 19, No. 6, March 1983, pp. 218–219.

[58] Drossos, G., Z. Wu, and L. E. Davis, "Switchable Cylindrical Dielectric Resonator Antenna," *Electronics Letters,* Vol. 32, No. 4, 1996, pp. 281–283.

[59] Kingsley, S. P., and S. G. O'Keefe, "Beam Steering and Monopole Processing of Probe-Fed Dielectric Resonator Antennas," *IEE Proc. Radar, Sonar and Navigation,* Vol. 146, No. 3, June 1999, pp. 121–125.

[60] Balanis, C., *Antenna Theory: Analysis and Design,* New York: John Wiley and Sons, 1997.

[61] Giger, A. J., *Low Angle Microwave Propagation: Physics and Modeling,* Norwood, MA: Artech House, 1991.

[62] Ruze, J., "Lateral Feed Displacement in a Paraboloid," *IEEE Trans. Antennas and Propagation,* Vol. 13, 1965, pp. 660–665.

[63] Durnan, G. J., D. V. Thiel, and S. G. O'Keefe, "Switched Parasitic Feeds for Parabolic Antenna Angle Diversity" *Microwave and Optical Tech. Letters,* Vol. 23, No. 4, 1999, pp. 200–203.

[64] Durnan, G. J., D. V. Thiel, and S. G. O'Keefe, "Optimization of Microwave Parabolic Antenna Systems Using Switched Parasitic Feed Structures," URSI National Science Meeting, Boulder, CO, January 4–8, 2000, p. 323.

5

Antenna Optimization

5.1   Introduction to Optimization Methods

The required specifications of an antenna are dependent on the application. For example, the antenna used in a cellular telephone base station may be required to provide adequate signal strength in a designated area, and have minimal signal strength in all other directions. In this case, the requirement may be for a large beamwidth and also a large front-to-back ratio. The impedance bandwidth must also be sufficient for the communications application. As another example, the antenna required for a radar system might require a very narrow beamwidth, low sidelobe levels, and a very narrow bandwidth.

The process of obtaining an antenna design from the specifications is a difficult task, because the number of variables required to completely specify an antenna can be quite large. For example, a designer might need an antenna with a particular beamwidth, bandwidth, and input impedance, and yet may wish to constrain the antenna to a maximum size and have it be constructed from particular materials. In Chapters 2 and 3 a number of methods for calculating the characteristics of an antenna from its physical description were presented. Antenna designers wish to take a set of required antenna performance characteristics and from them determine the physical size, shape, and number of elements for the antenna. In most cases, no simple method is available to move from the specification to the physical arrangement of the array. In some cases, there may be no antenna shape that meets the performance sought. It is possible, however, to move toward a compromise solution to the problem through the use of optimization techniques. The process of optimization begins with a basic antenna design, and a small number of parameters are varied within realistic manufacturing limits to obtain the best performance match between the desired specification and the outputs of the design.

As a first step, knowledge of the basic limitations of various antenna configurations is required. All optimization techniques require the antenna designer to have knowledge of the electrical and radiation performance of the basic antenna design. These become the features on which the optimization routine can act. This method of approach requires that the characteristics of each new antenna design be calculated and compared against the "ideal" performance required by the optimum specification. The antenna performance or specification can be described by m different quantities c_1, c_2, ..., c_m. For example, c_1 might represent the S_{11} impedance match, c_2 might represent the front-to-back ratio for the antenna, and so on. All are quantitative values describing the performance of the antenna. The c_1, c_2, ... properties are elements of the overall description of the antenna performance, and form the vector C. The optimal performance C_{opt} is the desired antenna performance description, and $C(i)$ is the output from the ith iteration of the optimization routine. Figure 5.1 illustrates the basic building blocks of all optimization routines where the solution of the next set of antenna parameters is of primary importance. Optimization techniques include sequential uniform sampling, the Monte Carlo method, the simplex method, the gradient method, genetic algorithm, and simulated annealing. The selection of the next set of design parameters from the previous set depends on the optimization routine implemented.

A directed optimization technique uses information from the previous models as a guide in the selection of the next set of design parameters. Sequential uniform sampling and the Monte Carlo method are nondirected optimization techniques.

Initially the antenna designer needs to specify the type of antenna structure that is likely to give satisfactory results, and then to set some parameters that remain fixed during the optimization process. In this discussion, the fixed parameters are represented by the symbols p_1, p_2, ..., which are elements of the vector P. For example, p_1 might be assigned to represent the number of wire elements, p_2 might represent the length of all of these elements, and so on. Optimization routines are most efficient when a large number of the design parameters have been fixed; however, this restricts the number of possible solutions. The performance of this basic antenna structure is then improved by allowing a limited number of parameters to be varied. These have been designated as variable design parameters $t_1, t_2, \ldots, t_n,$ and are represented by the vector T in Figure 5.1. The combination of P and T completely specifies the antenna structure. This design is then submitted to the solver and the effectiveness of the design D is calculated. The solver may be an analytical function or a numerical modeling algorithm. The output of the solver is the list of the values $d_1, d_2, \ldots, d_m,$ for the parameters upon which the antenna will be optimized. The quality of the fit is based on the modeled outputs D and the desired antenna specification D_{opt}. These are combined into a single valued cost function D, and the algorithm seeks to minimize the value of D. If the minimum value has not yet been achieved, then T is varied and the new antenna design is submitted to the solver. The process is repeated until the cost function minimum has been found. At this

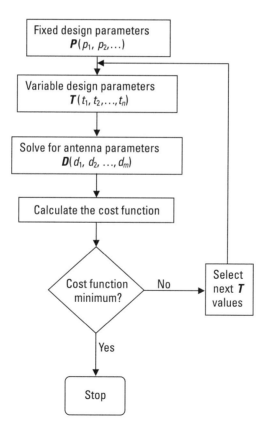

Figure 5.1 Basic building blocks of an optimization routine.

stage, the program is terminated and the complete antenna design is the combination of the fixed design parameters P and the variable design parameters T that give the lowest value of the cost function. Some of the processes in Figure 5.1 are discussed later in this section.

Optimization continues to be a challenging field. There are many emerging techniques as well as refinements of existing techniques which have been developed to improve the efficiency of the optimization process. While there are many different algorithms in use, the speed at which the best solution is found depends on the type of problem. An understanding of the effects of the various parameters on the optimal solution can guide the designer to the most efficient optimization technique. In this chapter, we discuss the basics of a number of optimization techniques, based on the assumption that the effect of the different variable design parameter T is not known.

With rapidly increasing computer speed, optimization routines that once were considered very slow might find favor in the years to come. It is important to compare the optimization time with the time required to build and test an antenna. In this light, an optimization routine that takes a day or more to provide a good solution might be a very efficient use of design time.

The choice of the fixed parameters P that have some likelihood of success requires a background knowledge of the material covered in Chapters 2, 3, and 4, in addition to some experience with the type of antenna structures that are used. If a suitable design cannot be found from the optimization process, it is necessary to restart the optimization process with a new set of fixed parameters P.

5.1.1   Cost Function

An important parameter in the optimization routine is the definition of the cost function. This is a single number used to quantify how well a design meets the specifications. It is also known as the error function or the objective function. It is used to compare the desired characteristics of the antenna with those of all previous iterations. For example, if the desired antenna characteristics D_{opt} include a front-to-back ratio of FB_{opt} dB, a 10-dB beamwidth of $\phi_{opt}°$, and an S_{11} of S_{opt} dB, the cost function must involve contributions from all of these parameters. One common approach is to define the cost function in terms of the least squared error. If the parameters calculated from the ith iteration of the optimization loop are the front-to-back ratio of $FB(i)$ dB, the beamwidth of $\phi(i)°$, and the $S_{11} = S(i)$ dB, then the least squared error cost function $C_{lse}(i)$ is given by:

$$C_{lse}(i) = \sum (D_{opt} - D(i))^2 = (FB_{opt} - FB(i))^2 + (\phi_{opt} - \phi(i))^2 + (S_{opt} - S(i))^2$$
(5.1)

The object of the optimization loop is to continue to change the variables in the antenna model until the minimum value of C_{lse} is reached. In principle, the cost function given by (5.1) gives a reliable estimate of the closeness of the specification parameters to those of the latest model.

There are weaknesses is this approach. If one of the optimum values is very large in comparison with the others, then it will dominate C_{lse} and the antenna will be optimized preferentially for this performance indicator. Similarly, if one parameter is very small, it will be ignored in the optimization process. An improved cost function, the relative least squared error $C_{rlse}(i)$, assigns equal weights to all parameters and is given by:

$$C_{rlse}(i) = \left(\frac{FB_{opt} - FB(i)}{FB_{opt}}\right)^2 + \left(\frac{\phi_{opt} - \phi(i)}{\phi_{opt}}\right)^2 + \left(\frac{S_{opt} - S(i)}{S_{opt}}\right)^2$$
(5.2)

Even with (5.2), the influence of one parameter may be more significant in C_{rlse}. For example, a 1% error in one parameter may cause a change in the cost function that is much larger than the change caused by a 50% variation in another parameter. The individual components in (5.2) can be weighted unevenly to protect the system from one dominating factor or bias. In general, the weighted cost function for the ith iteration in the optimization procedure can be written:

$$C_{rlse}(i) = \sum_{j=1}^{m} w_j \left(\frac{d_{j,opt} - d_j(i)}{d_{j,opt}}\right)^2$$
(5.3)

where $d_{j,opt}$ is the desired specification value, $d_j(i)$ is the calculated value of the jth parameter, w_j is the weighting factor for the jth parameter, and m is the total number of parameters used in defining the cost function.

In many cases it is desirable to minimize various parameters rather than to seek an optimal value. For example, if the best performance of an antenna is to have $FB_{opt} = \infty$ dB and $S_{opt} = -\infty$ dB, these values will never be reached, and so FB_{opt} should be maximized and S_{opt} should be minimized. If the total cost function $C_{tot}(i)$ has s unconstrained parameters to be minimized and $(m-s)$ constrained parameters to be optimized, then:

$$C_{rlse}(i) = \sum_{j=1}^{m-s} w_j \left(\frac{d_{j,opt} - d_j(i)}{d_{j,opt}} \right)^2 + \sum_{k=1}^{s} w_k d_k(i) \qquad (5.4)$$

The weights for the unconstrained parameters w_k may be quite difficult to determine, and the overall balance of the weights requires intuition and often some trial and error. If the solver is quite slow, then it is usually best to use C_{rlse} as defined in (5.3) by making the unconstrained parameters constrained to a desirable level. For example, if $FB_{opt} > 20$ dB, then when $d_k = FB > 20$ dB, this factor can be removed from the cost function calculation by setting the weighting $w_k = 0$.

In some circumstances, one of the terms in the cost function may vary slowly while the other terms vary rapidly. This will tend to focus the efforts of the optimization process on the rapidly varying term only. The final result may not be the lowest value of the cost function; rather, a local minimum in the cost function will be discovered. A better solution with a much lower cost function and a greatly improved antenna performance may be missed. The solution of this problem is addressed in different ways by a number of different algorithms used to update the variable design parameters in the loop. These techniques include sequential uniform sampling, the Monte Carlo method, the simplex method, the gradient method, the genetic algorithm method, and the simulated annealing method. Each of these techniques is introduced in a simple form in the following sections of this chapter. It must be emphasized that there are many variations to these techniques. A basic introduction to each one is given here.

It is possible that the definition of the cost function may result in an unusual antenna performance. For example, the optimization process may yield an antenna design with a very small cost function, but one that is unsuited to the required application. For example, it may be that there are very large sidelobes that are not directed at the $\phi = 180°$ position. The front-to-back ratio requirement may be satisfied, but the sidelobes might mean that the antenna is unsuitable. One solution to these problems is to include additional terms in the cost function, limiting unwanted performance characteristics, or to weight the set of optimum parameters differently.

5.1.2 Computational Techniques

The antenna solver is used many times during the optimization process. The method of solution should be extremely efficient so that a solution is possible within a reasonable time frame. If this step takes several hours, then the time required to optimize a structure through many design iterations may be days or weeks. If the solver works in the frequency domain and bandwidth information is required, each pass of the optimization loop will require more than one calculation of the antenna characteristics. For example, if the desired bandwidth is $2f_b$, around the center frequency f_0, then each optimization step requires a minimum of two calculations. In this case it may be beneficial to conduct the calculation at three different frequencies—f_0, $(f_0 + f_b)$, and $(f_0 - f_b)$—if the gain or front-to-back ratio of the antenna is part of the specification.

Another constraint on the optimization process is the accuracy of the computation. It is sometimes efficient to use a high-speed, but approximate, computational technique to obtain an approximate solution, and then to use a more accurate but slower solver for the final optimization. For wire antennas, the NEC code [1] is both efficient and accurate for most situations. There are approximations involved when modeling ground planes of finite size or ground planes that contain bends. In these situations, the efficiency of the optimization process begins to fade. In such cases, it may be necessary to use an infinite ground plane in the initial optimization routine, and once a coarse solution has been achieved, to move to a more accurate technique. For patch antennas, full wave solvers may be less efficient, and so alternative solution methods are required [2, 3]. For complex problems involving a combination of metallic and lossy dielectric materials, a full solution of Maxwell's equations may be required, using numerical codes such as the finite-difference time-domain (FDTD) method [4, 5] or the finite-element method [6]. While there are many commercial electromagnetic solvers available, the executable (*.exe) files in a batch input/output mode are required in the process of optimization. Electromagnetic solvers without this type of access cannot be used easily; however, many computer programs available for solving EM problems can now be purchased with companion optimization routines [6].

The efficiency of the various optimization techniques can be measured by the number of times a new design is submitted to the solver. This will be referred to as the iteration number. This should not be confused with the number of times the loop shown in Figure 5.1 is run. Some optimization techniques require more than one call to the solver for each loop in the optimization algorithm.

5.1.3 Sensitivity Analysis

The final output of the optimization process is an antenna design. The manufacturing process to construct the antenna requires an understanding of the tolerances in the mechanical design. In the general structure of the optimization process, it is usually necessary to search for the optimum mechanical configuration by locating the global minimum in the cost function. The sensitivity of that global minimum value to small changes in the mechanical design is a measure of the manufacturability of the antenna. It is best to have a relatively small change in the cost function in the vicinity of the global minimum, because this means that small variations in the mechanical construction will not have a significant effect on the overall performance of the antenna.

An outcome of the nondirected methods of searching for the cost function minimum is that the data can be used in a sensitivity analysis (sometimes referred to as a parametric analysis). This is the change in the cost function as a function of the change in one or more specific design parameters. The more sophisticated optimization techniques may not provide this information, and additional investigations must be undertaken. Given the physical limitations of a particular construction technique, it is not useful to discretize the parameters to be optimized to an accuracy that is finer than can be realized physically. This can be used as a guide to the level of discretization to be used in the optimization process. It is important to record the history of the optimization process, which includes the iteration number, the input parameters, and the cost function. A plot of the input parameters versus the cost function is an effective tool in locating the best design as well as in assessing the sensitivity of that design to manufacturing variability.

5.1.4   Example: The Three-Element Reflector Antenna

In order to illustrate how the various optimization techniques can be implemented, a relatively simple design will be used as an example. This antenna is required to provide 180° coverage in the H plane with a large front-to-back ratio and a nominal 50Ω input impedance. In this way, almost omnidirectional coverage is achieved, but the effect of a strong interfering source behind the antenna is reduced by positioning the radiation pattern null in this direction. The structure chosen to provide this performance is the three-element dipole antenna shown in Figure 5.2. As with most antennas, a design with the smallest possible volume that meets the specifications is sought.

For simplicity, it is assumed that the elements are all half-wave dipoles with radius $0.0001\lambda_0$, oriented parallel to the z axis with the center of each element located in the xy plane, which also defines the H plane of the antenna. The driven element (1) is located at a distance t_1 from the yz plane containing the two parasitic elements (2) and (3). The separation distance

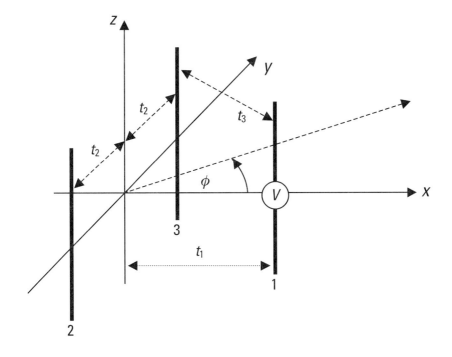

Figure 5.2 Three-element dipole antenna used to test optimization procedures. Element 1 is driven, and elements 2 and 3 are parasitic elements. The angle ϕ defines the H plane radiation pattern. The distance between elements 1 and 2 and 1 and 3 is t_3, and the distance between elements 2 and 3 is $2t_2$.

between the x axis and the parasitic elements is t_2. In this design, t_1 and t_2 are optimized simultaneously. The three performance parameters used in the cost function are $FB_{opt} = 25$ dB, a gain of -3 dB at $\phi_{opt} = \pm 90°$ measured relative to 0 dB at $\phi_{max} = 0°$, and $S_{opt} = -20$ dB.

The analytical method used to calculate the radiation pattern for such an antenna, discussed in Chapter 2, does not yield a direct measure of the beamwidth. While this can be done using an interpolation technique, it is far simpler to optimize the radiation pattern for the ϕ values, given in Table 5.1, referenced to the angle of the main beam ϕ_{max}.

If the antenna elements remain in the same plane, the radiation pattern is symmetrical about the x axis. Therefore, we can write:

$$BW_{opt} = BW1_{opt} = BW2_{opt} \tag{5.5}$$

Table 5.1
Optimal Specification for the Three-Dipole Antenna

Label	Angle ϕ (degrees)	Gain (dB)
ϕ_{max}	0	0
$BW1_{opt}$	90	-3
FB_{opt}	180	-25
$BW2_{opt}$	270	-3

In order to minimize the computation time, the gain at three angles only are calculated, that is, for $\phi = 0°$, 90°, and 180°.

The cost function becomes:

$$C_{rlse}(i) = \frac{4}{6}\left(\frac{FB_{opt} - FB(i)}{FB_{opt}}\right)^2 + \frac{1}{6}\left(\frac{BW_{opt} - BW(i)}{BW_{opt}}\right)^2 + \frac{1}{6}\left(\frac{S_{opt} - S(i)}{S_{opt}}\right)^2 \quad (5.6)$$

The weights have been chosen to emphasize the front-to-back ratio.

For computational speed, a nine-term polynomial fit is used to approximate the mutual impedance curves for two half-wave dipoles (radius $0.0001\lambda_0$) given in Figure 2.8. The variations in the real and imaginary parts of the mutual impedance $R(t)$ and $X(t)$, respectively, are given as a function of the separation distance t:

$$R(t) = \sum_{i=0}^{8} a_i t^i \quad (5.7)$$

$$X(t) = \sum_{i=0}^{8} b_i t^i \quad (5.8)$$

and the coefficients a_i and b_i are given in Table 5.2, where the impedances $R(t)$ and $X(t)$ in (5.7) and (5.8) are given in ohms.

Equations (5.7) and (5.8) are used to calculate the mutual impedance between the two dipoles separated by a distance t. This technique is used as the solver in the optimization routine, as it is more efficient than using either the MOM code or the induced EMF method. The mutual impedance elements are part of the 3×3 impedance matrix used in (2.38) to calculate the

Table 5.2
Polynomial Coefficients for the Calculation of Mutual Impedance Using (5.7) and (5.8)

i	0	1	2	3	4	5	6	7	8
a_i	73	7.9	−576.4	−834.7	4,968.6	−6,643	4,137.2	−1,302.3	173.4
b_i	44	270	−1,422	8,951	−19,590	23,752	−17,076	6,746	−1,117

current in each element. The relative field strength in the far field is calculated as a function of the angle ϕ from the currents in each element.

From Section 2.4, the mutual coupling is significant when the spacing between adjacent dipole elements is less than approximately λ_0. For this reason t_1 and t_2 are chosen to be less than $0.64 \lambda_0$, so then $t_3 < \lambda_0$. Following the discussion given in Section 5.1.3, the two parameters are varied over the ranges:

$$0.01\lambda_0 \le t_1 \le 0.64\lambda_0 \quad \text{and} \quad 0.01\lambda_0 \le t_2 \le 0.64\lambda_0 \qquad (5.9)$$

and so the element spacings $2t_2$ and t_3 vary over the ranges:

$$0.02\lambda_0 \le 2t_2 \le 0.128\lambda_0 \quad \text{and} \quad 0.02\lambda_0 \le t_3 \le 0.9\lambda_0 \qquad (5.10)$$

This example is used to demonstrate how a number of optimization techniques can be implemented. In Section 5.8, each of the methods discussed is reviewed for efficiency in obtaining the cost function minimum.

5.2 Sequential Uniform Sampling

Sequential uniform sampling requires the cost function to be calculated for all possible values of the input variables in the solution space; in this example, these are t_1 and t_2. The resolution is determined by the step size used in the input parameters. This determines the number of iterations required for full coverage of all possible input combinations. This is similar to the uniform scanning procedure used to locate the angular position of an unknown signal source described in Chapter 1.

In the case of the problem outlined in Section 5.1.4, the two distances t_1 and t_2 would be sequentially changed and applied to the solver. If t_1 is varied through r possible values and t_2 is varied through s possible values, the value for the ith iteration of t_1 and the jth iteration of t_2 can be written as:

$$t_1(i) = t_{1\,min} + r(i)\Delta t_1$$
$$t_2(j) = t_{2\,min} + s(j)\Delta t_2 \qquad (5.11)$$

where the iteration numbers $r(i) = 0 \ldots r$, and $s(j) = 0 \ldots s$, t_{1min} and t_{2min} are the minimum values, and Δt_1 and Δt_2 are the iteration steps of the distances t_1 and t_2, respectively. For illustration here, $\Delta t_1 = \Delta t_2 = 0.01\lambda_0$ has been chosen. There are 64 steps for both t_1 and t_2 in the ranges specified in (5.9). This requires a total of 4,096 combinations to be solved.

It is also possible to represent the variable input vector T as a single binary number. In the genetic algorithm method discussed in Section 5.6, this number is identified as a chromosome. If both t_1 and t_2 run through 64 possible values independently (i.e., $r = s = 64$), then the binary number will have 12 digits. For example, the binary number:

$$T = 001001\ 100010 \qquad (5.12)$$

represents $r_i = 9$ and $s_i = 34$ corresponding to the 609th iteration (note that there is a zeroth step) in the total sequence of 4,096 steps ($= 2^{12}$).

In the sequential uniform sampling method, this binary (or chromosomal) representation of the total input variables must be iterated from 0 to 4,095 sequentially, and the cost function determined for each model. The minimum value of the cost function determines the best set of design parameters T. Higher resolution can be obtained if this cycle is repeated using a more localized search with smaller values of Δt in the vicinity of the cost function minimum. The initial search must be sufficiently fine in resolution to ensure that the global minimum value of the cost function is found.

Figure 5.3(a) shows a surface plot of the cost function C_{rlse} as both t_1 and t_2 are varied sequentially. The C_{rlse} surface is undulating, and there are a number of possible minimum values. While the global minimum can be seen in this plot, it is more apparent as a maximum value in a surface plot of the reciprocal value $1/C_{rlse}$ given in Figure 5.3(b). In this plot there are a number of maximum values, although one is significantly larger than the others. Note that if the variable design parameter T contains more than two variables, then a surface plot representation is not possible. The nonvisual multidimensional abstraction of the cost function surface is necessary. The mathematical treatment, however, is identical.

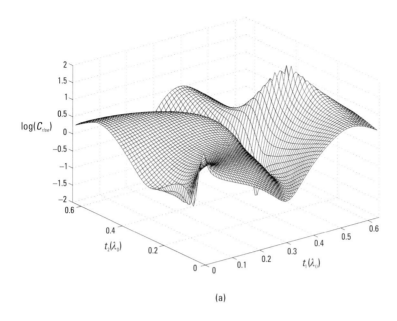

(a)

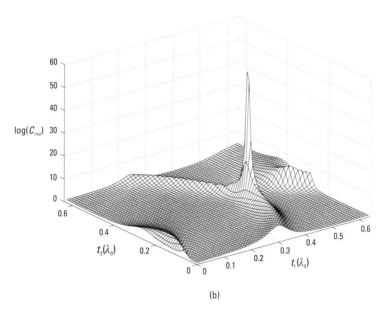

(b)

Figure 5.3 (a) Variation in the cost function C_{rise} as a function of t_1 and t_2. (b) Variation in $1/C_{rise}$ as a function of t_1 and t_2. This data was determined from the sequential uniform sampling technique described in Section 5.2. The optimized values of t_1 and t_2 coincide with the minimum value of C_{rise} and the maximum value of $1/C_{rise}$.

An alternative view of the same data is given as a scatter plot in Figure 5.4, where the values of t_1 and t_2 are plotted as a function of C_{rlse}. For every value of C_{rlse}, there is one value of t_1 and another for t_2. The minimum value of C_{rlse} corresponds to $t_1 = 0.42\lambda_0$ and $t_2 = 0.31\lambda_0$. While this is the most significant C_{rlse} minimum, the undulating nature of the surface suggests that there may be other minima at larger values of t_1 and t_2. Normally the strongest minimum value offered by the optimization routine is chosen, although it is tempting to look at other possible solutions at larger values of t_1 and t_2 that were not initially modeled. In this case, however, the cost function minimum located in Figure 5.3 has the smallest antenna size. This is usually a desirable property of any antenna. The final decision must involve a more detailed look at other possible cost function minima, their associated radiation patterns, and the manufacturing tolerance requirements.

One advantage of this technique is the ability to locate a number of minimum values for the cost function, which offers some flexibility in the chosen solution. Many of the optimization techniques discussed in this chapter do not offer this possibility unless the search space is divided into a

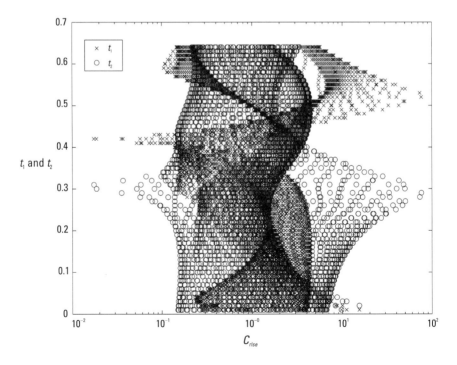

Figure 5.4 Variation in t_1 and t_2 plotted as a function of C_{rlse}.

number of different ranges, and each search is conducted independently. A major disadvantage is the large number of calculations required.

Figure 5.4 also gives an indication of the sensitivity of the optimal antenna design to variations in t_1 and t_2 that might result from manufacturing inaccuracies. The difference in cost function close to the maximum is quite large. These adjacent points are on the shoulders of the minimum values and are removed from the large majority of the data values. For this design, the manufacturer would be required to work within a mechanical tolerance of $\Delta t_1 = \Delta t_2 = 0.01\lambda_0$ so that antenna performance is not affected.

The final design of the antenna is summarized in Table 5.3, together with the performance parameters initially specified in the cost function. The smallest cost function in the solution space defined by (5.9) is $C_{rlse} = 0.017$ for $t_1 = 0.42 \lambda_0$ and $t_2 = 0.31 \lambda_0$. For an independent comparison, the final design was verified using NEC [7]. The results are listed in Table 5.3 and conform quite well to the initial specification.

The cost function C_{rlse} has also been plotted as a function of the iteration number in Figure 5.5. In this method, the number of iterations required is known from the discretization of the parameters. In other methods, the number of iterations may be unknown initially, and will depend on the speed of convergence of the cost function to the global minimum. Figure 5.5 does not give an indication of the efficiency of the optimization routine, as the start position given in (5.11) contains the two lowest values. At iteration 200, a local minimum occurs. If the routine had tracked the minimum cost function value, this may have been identified as the global minimum, and the routine would have stopped. The global minimum at iteration 2,600 would then have been missed.

The H plane radiation pattern of this antenna is given in Figure 5.6 with the almost identical NEC result. The input impedance calculated from

Table 5.3
Results for the Sequential Uniform Search Optimization for the Design Task Outlined in Section 5.1.4 *

	FB_{opt}	BW_{opt}	S_{opt}
Polynomial result	20.9 dB	−2.9 dB	−14.6 dB
NEC result	20.8 dB	−2.96 dB	−13.5 dB
Optimization target	25 dB	−3 dB	−20 dB

* Calculated using the polynomial function compared with the NEC calculation and the initial specification.

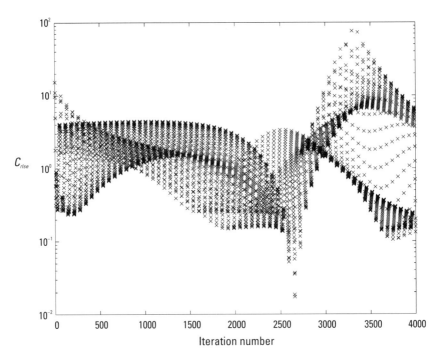

Figure 5.5 Variation of C_{rlse} as a function of iteration number of the sequential uniform sampling method.

the optimization routine is $69 + 12j\,\Omega$. This is quite close to the NEC [7] result of $74 + 11j\,\Omega$. If one of the specifications in the cost function is to be emphasized in comparison with the others, then the weighting factors given in (5.6) can be altered and the design can be reoptimized.

5.3   The Monte Carlo Method

The Monte Carlo method of solving complex problems involves the use of random numbers as inputs to the problem, whether the application is the evaluation of an integral or involves tracking the path of a neutron through a matrix of atoms [8]. The name arose from the gambling tables at the casinos of Monte Carlo. In antenna optimization, the method is based on the generation of random numbers that make up the variable input values T of the optimization routine shown in Figure 5.1. It is also known as the blind random search method [9]. As before, the design task from Section 5.1.4 is used

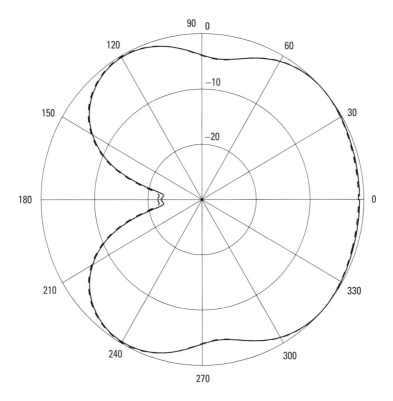

Figure 5.6 H plane radiation pattern (normalized dB) of the optimized three-element antenna compared with the NEC result for the same array.

to demonstrate how the Monte Carlo method is used to optimize the three-element dipole antenna shown in Figure 5.3. The cost function C_{rlse} given by (5.7) and the ranges of the variable parameters t_1 and t_2 in (5.10) are used.

In the Monte Carlo method, t_1 and t_2 are determined from a random number generator and lie somewhere between the upper and lower bounds. As with the sequential uniform sampling technique, the method is not guided to the required solution. Rather, it is anticipated that the parameters selected will eventually cover the whole of the parameter space randomly. In this way a set of cost function values are determined from a bounded but random set of input parameters.

The ith iteration of the two variables can be written as:

$$t_1(i) = t_{1\,min} + rand(r)\Delta t_1$$
$$t_2(i) = t_{2\,min} + rand(s)\Delta t_2 \tag{5.13}$$

where *rand*(r) is the random number function that returns an integer number in the range 0 to r. Most high-level computer programming languages provide a pseudorandom number function that returns a number within the range of zero and one. This number is multiplied by r or s and the integer value of the result is used in (5.13).

In the chromosome representation of the input variables described by (5.12), the Monte Carlo method requires the random selection of the binary number, which represents all of the variable input parameters T. The integer value of the output from a random number generator is multiplied by the total number of bits to give rs.

As before, the results can be analyzed by plotting the parameter values on the y axis and the cost function on the x axis. Figure 5.7 illustrates results for a 400-point Monte Carlo optimization for the three-element dipole antenna. The minimum value in the cost function is apparent by one point only, but of course, with random statistics, results will change for every run. The probability of missing the global minimum is related to the number of points used in the Monte Carlo simulation. If the minimum value in the cost function is very sharp, then the global minimum might be missed. As indicated earlier, this is also related to the manufacturability of the antenna and the step sizes used for the parameters in (5.13).

Figure 5.8 shows the variation in the cost function with iteration number. As the choice of values of t_1 and t_2 is always random, the one point that is a clear minimum in this plot will not appear in every run of the algorithm.

It is evident from Figures 5.7 and 5.8, that the Monte Carlo method can provide a good guide to the location of the position of the cost function minimum. Once this has been performed, a follow-up in the general area of the minimum can be made using the sequential uniform sampling method or a surface fitting routine on the data acquired. This will give a stronger indication of the location of the minimum value.

In this run, approximately 10% of the points needed for the sequential uniform sampling method were used. A minimum in the cost function has been identified, but further work is required to resolve the position within the resolution of the step size used. A point worthy of further discussion is whether the Monte Carlo method offers an advantage over the sequential uniform sampling technique, with a coarse grid requiring the same number of sample point calculations. The advantage of the Monte Carlo method is that the nonuniform sampling may provide information that is not found in a uniform grid, should the cost function surface be irregular.

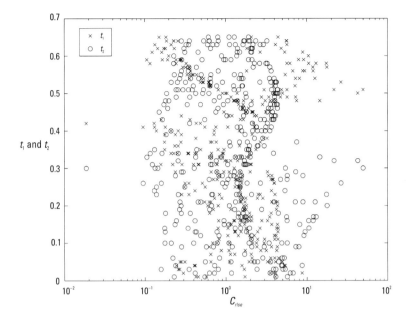

Figure 5.7 Results from a 400-point Monte Carlo method for the three-element dipole antenna in Figure 5.2.

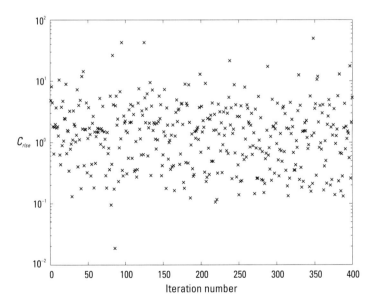

Figure 5.8 Plot of the cost function as a function of the iteration number for the 400-point Monte Carlo method.

5.4　The Simplex Method

The simplex method is a relatively simple, directed search strategy. It is in a class of guided search methods used for multiparameter optimization [9], where the value of T chosen for the next iteration of the optimization routine is derived from the results from the previous iteration.

Unlike the gradient method discussed later, the simplex method does not require the calculation of any derivatives of the function to be minimized. It does, however, require the calculation of the cost function at adjacent points in the solution space, called a simplex. The first point is selected randomly. If n is the number of variables in T, then the first step is to calculate $(n + 1)$ values of the cost function at positions adjacent to the first point. In its simplest form, these points are separated by one incremental step in the direction of each parameter [9–12]. In the example given in Section 5.1.4, there are two variable parameters t_1 and t_2, and so the number of initial calculations required is three. The simplex is a triangle on the cost function surface.

The search progresses using three processes (see Figure 5.9). The update equation used first is referred to as reflection. The point with the highest cost function is reflected in the centroid $T_c(i)$, calculated from all points except the one to be replaced. $T(i)$ is replaced by $T(i + 1)$ using:

$$T(i+1) = (1+\alpha)T_c(i) - \alpha T(i) \tag{5.14}$$

where α is called the reflection coefficient and:

$$T_c(i) = \frac{1}{n}\sum_{\substack{j=1 \\ j \neq i}}^{n+1} T_j(i) \tag{5.15}$$

If $\alpha = 1$, then the new point $T(i + 1)$ is equidistant from $T_c(i)$ but in the opposite direction (see Figure 5.10). If the new point $T(i + 1)$ has the highest cost function value in the next simplex, it will be reflected back to $T(i)$, creating an infinite loop. This can be avoided by using a slightly smaller value such as $\alpha = 0.9985$.

If $T(i + 1)$ has a significantly lower cost function than $T(i)$, then the reflection operation has $T(i + 1)$ closer to the minimum value of the cost function. The step in this direction can be extended beyond the reflection point but still located along the same line. This operation is called expansion and is used when the cost function of the new point $T(i + 1)$ calculated from

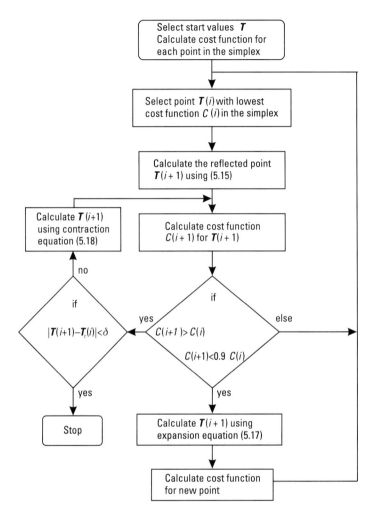

Figure 5.9 Simplex method algorithm to calculate next vector $T(i + 1)$ from the highest cost function value in the simplex structure from the previous set, $T(i)$.

(5.14) indicates that the search is moving in the correct direction with some certainty (see Figure 5.10). This process is given by the equation:

$$T(i+1) = \beta T(i+1) + (1-\beta)T_c(i) \qquad (5.16)$$

where β is the expansion coefficient. If $\beta = 2$, then the new point has moved twice as far from the centroid as the initial reflection operation.

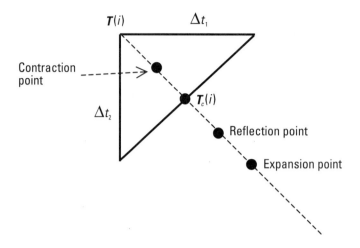

Figure 5.10 Two-dimensional simplex operations of reflection ($\alpha = 1$), expansion ($\beta = 2$) and contraction ($\gamma = 0.5$) showing the location of the next points along the line joining the old point $T(i)$ and the centroid $T_c(i)$.

If the cost function of the reflected point is greater than that from the other points in the simplex, then the step distance is too large. The step size is reduced by a contraction operation defined by:

$$T(i+1) = \gamma\, T(i) + (1 - \gamma)T_c(i) \tag{5.17}$$

where γ is the contraction coefficient. For a step size of one-half the reflection operation, $\gamma = 0.5$. The contraction operation is repeated with $\gamma \to 1$ until either the cost function value is smaller or the step size is so small as to be insignificant. Thus, when

$$\left|T(i+1) - T_c(i)\right| < \delta \tag{5.18}$$

the routine is terminated. Here δ is defined as the step size that is so small as to have an insignificant effect on the cost function.

The decision about which process to select depends on a comparison between the cost function $T(i + 1)$ of the new point compared with that from the point rejected after reflection $T(i)$. The coefficients are selected to be approximately [9]:

$$\alpha = 0.9985 \quad \beta = 1.98 \quad \gamma = 0.498 \tag{5.19}$$

Noninteger values are chosen, to prevent an infinite loop from forming, while still maintaining the stability of the process.

The path through the solution space is a zigzag progression toward a minimum value in the cost function. A minimum value is identified when the inequality in (5.18) is no longer true. Here the algorithm becomes permanently trapped and any more iterations yield the same cost function result

The simplex method is usually trapped at the first minimum that is encountered. This might be a local minimum rather than the global minimum. For this reason, the selection of the starting location should be randomized using (5.13) and the algorithm run a number of times. It is necessary to compare the results from each of the runs. After a number of runs, the lowest value of C_{rlse} should correspond to the global minimum.

There are many variations to the simplex method that have been developed to increase the speed of convergence and to avoid termination in a local mimima. These enhanced techniques are not discussed here, but further information is available in the literature [10–12].

The simplex method was applied to the design task given in Section 5.1.4. The results from four runs with different, randomly selected initial values are given in Figure 5.11, where the variation in t_1 and t_2 is traced through the solution space. Only one path successfully located the global minimum at $t_1 = 0.42\lambda_0$ and $t_2 = 0.31\lambda_0$. The other three paths were trapped at local minima. The zigzag path is evident in the figure as the iterations progress. The successful path was trapped for some time on a trough of minimum values resulting in a quite indirect path to the optimum solution. Figure 5.12 shows the variation in C_{rlse} as a function of iteration number. The convergence to the final result is evident. As expected, the speed of convergence depends on the start value. The oscillations in the successful trace show where the infinite loop condition was approached, but was eventually resolved. The contraction operation is the final operation of the technique as the inequality in (5.18) is reached. This means that the resolution of the technique is not confined to Δt_1 and Δt_2 values of the initial simplex, but is determined by the factor δ.

5.5 The Gradient Method

The method of steepest descent is a multiparameter optimization technique in which the search is always in the direction of the gradient of the cost function $C(T)$ [9–11]. The gradient G is a vector defined in terms of the partial

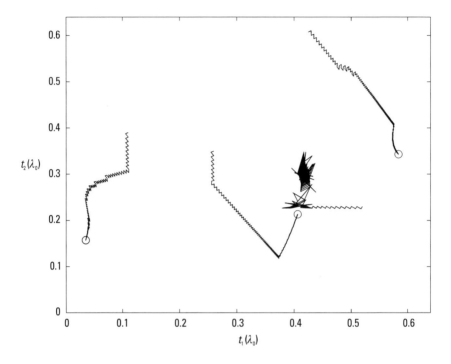

Figure 5.11 Simplex search path for four randomly selected start points. The final point in each run is marked by an o. Only one of the four start positions located the global minimum value at $t_1 = 0.42\lambda_0$, $t_2 = 0.31\lambda_0$.

derivatives of the function with respect to all components individually. If n is the number of independent variable design parameters, then:

$$G = \nabla C(T) = \frac{\partial C}{\partial t_1}, \frac{\partial C}{\partial t_2}, \frac{\partial C}{\partial t_3} \dots, \frac{\partial C}{\partial t_n} \qquad (5.20)$$

The method is initialized by the random selection of T using (5.14). The cost function C_{rlse} is calculated for this value and all adjacent values in the $+t_r$ directions, where $r = 1 \dots n$. This gives sufficient information for the discrete calculation of the components of the gradient G. The iterative process to find the minimum value of the cost function can be written as:

$$T(i+1) = T(i) - \lambda(i)G(T(i)) \qquad (5.21)$$

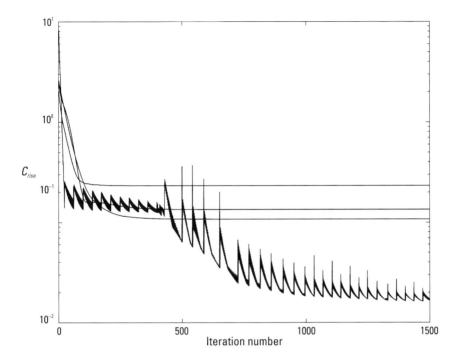

Figure 5.12 Variation in the cost function as a function of iteration number using the simplex method for four independent trials. The start points were randomly selected. Only one of the four runs located the global minimum.

where $\lambda(i)$ is a constant that determines the step size. If the step size is too large, then the algorithm is unstable, and if it is too small the convergence is very slow. One method of determining the step size is to relate it to the construction limitations of the antenna. If the step size of the rth parameter in T is Δt_r, then $\lambda(i)$ can be calculated from the equation:

$$\lambda(i) = \frac{\Delta t_r}{\left| \dfrac{G(T(i))}{\Delta t_r} \right|_{max}} \tag{5.22}$$

If the cost function for the new $T(i + 1)$ is larger than that for $T(i)$, then $\lambda(i)$ is replaced with $\lambda(i)/2$ and the cost function is recalculated. This preserves the slope of the gradient function but reduces the magnitude of the step. This process is continued until $\lambda(i)$ is so small that no change in the cost function will result. This minimum value of $\lambda(i)$ is shown as δ in Figure 5.13. This means that the minimum value of the cost function has been located and the algorithm is terminated. The complete process is illustrated in Figure 5.13.

This process was applied to the design task in Section 5.1.4 and the results are plotted in Figures 5.14 and 5.15. Four randomly selected start positions were used, but only one was successful in locating the minimum value of the cost function. While the number of iterations in (5.21) was relatively small, the total number of calls to the solver required to vary the value of the $\lambda(i)$ was much larger. In this case the number of iterations required for convergence was of the order of 100. Figure 5.14 shows the path taken through the solution space to find a minimum value, and Figure 5.15 shows the variation in the cost function with the iteration number. The sporadic increases in the cost function evident in Figure 5.14 results from the procedure to define

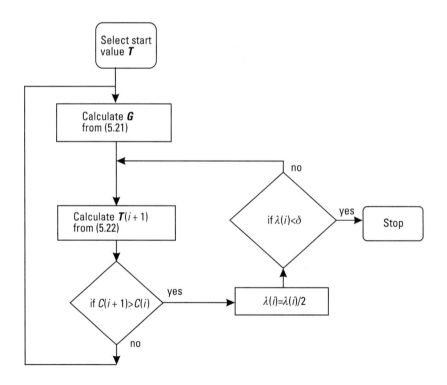

Figure 5.13 Gradient algorithm to calculate the next vector $T(i+1)$ from the gradient of the previous point $G(T(i))$.

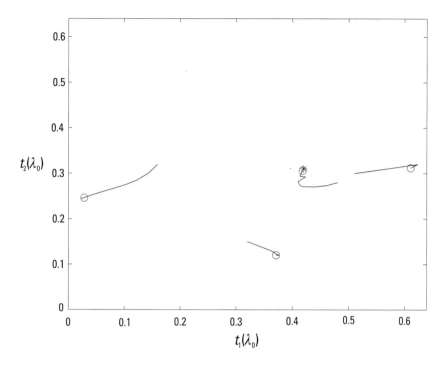

Figure 5.14 Schematic diagram showing four different runs of the gradient method. The terminating value of **T** is marked o. Only one run located the minimum cost function value at $t_1 = 0.42\lambda_0$ and $t_2 = 0.31\lambda_0$.

the gradient *G* given by (5.20). By definition, the updated position $T(i + 1)$ always results in a cost function less than the previous one.

The gradient method is efficient in locating a minimum value of the cost function, but the result is often a local minimum rather than the global minimum.

5.6 Genetic Algorithm

Optimization routines based on the genetic algorithm (GA) have recently become popular in electromagnetics [13–20]. The initial concept has been attributed to Goldberg [21]. The process imitates biological reproduction, where information about an individual is coded into a DNA strand. In reproduction, the strands from two genetically different parents are brought together and the resulting children carry information derived from both

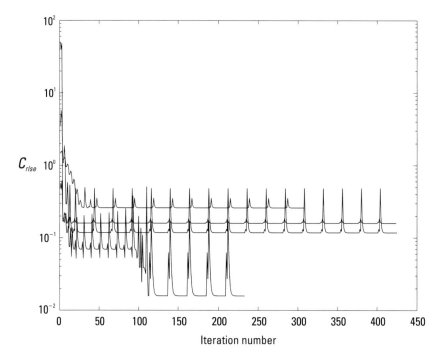

Figure 5.15 Speed of convergence of the gradient method is illustrated as the cost function varies with the iteration number for four different start values of *T*.

parents and perhaps some randomized alterations in the DNA structure. Following Darwin's theory of natural selection [22], only those children with desirable attributes survive and go on to parent the next generation. If only the strongest are allowed to survive, it is assumed that the overall characteristics of the next generation will be better than the last.

In the process of optimization outlined in Section 5.1, the variable input parameters T are described by a single binary coded number called a chromosome, (5.12). The various elements of T, t_1, t_2, ..., t_n, are the "genes," or parts of the chromosome that represent particular properties of the antenna. Each t value is represented by a finite number of binary digits (bits) that governs the resolution of the method as shown in (5.12). The optimization is conducted on a number of chromosomes simultaneously, referred to as the population p.

The chromosomes in the first population are usually created randomly using (5.13) and represented by a binary number, such as (5.12). The cost function for all members of the population is calculated. Those members of the population that are unsuitable (for example, the 50% with the highest cost function) are either discarded or assigned a low probability of being selected as a parent for the next generation. The latter technique is described as roulette wheel (or tournament) selection, where the probability of selection depends on the rank ordering of the population in terms of their cost function, and the subsequent allocation of probabilities related to their cost function values. There are four common methods of generating the next population:

1. Crossover;

2. Mutation;

3. Creation;

4. Cloning.

The balance between the four techniques in the next generation depends on the problem under consideration and the criteria used to terminate the optimization loop.

These are illustrated in Figure 5.16. The procedure for the ith generation for a single point crossover requires two parents $T_1(i)$ and $T_2(i)$, and a randomly selected cut point. The $(i + 1)$th generation consists of two offspring: $T_1(i + 1)$ formed from the left side of the first parent chromosome and the right side of the other, and $T_2(i + 1)$ formed from the left side of the second and the right side of the first. It is possible to use more than one crossover point, where the outer two pieces of the chromosome remain, with the center section being taken from another chromosome. It is not necessary that the cut point be located at a gene boundary in the chromosome, as this increases the genetic diversity of the population.

Mutation requires the arbitrary selection of a chromosome from the ith generation $T(i)$ and replacement by $T(i + 1)$ in which one or more bits in $T(i)$, selected at random, have been inverted (i.e., $1 \to 0$ or $0 \to 1$).

Creation is identical to the initial selection procedure using (5.13), with an entirely new chromosome being inserted into the population. Mutation and creation both allow the population to escape from a local minimum in the cost function, but with the consequential effect that there may be a continuing variation in the mean value of the population cost function. Cloning is a direct move of a chromosome from one generation to the next; that is, $T(i + 1) = T(i)$. These four techniques are illustrated graphically in Figure 5.16.

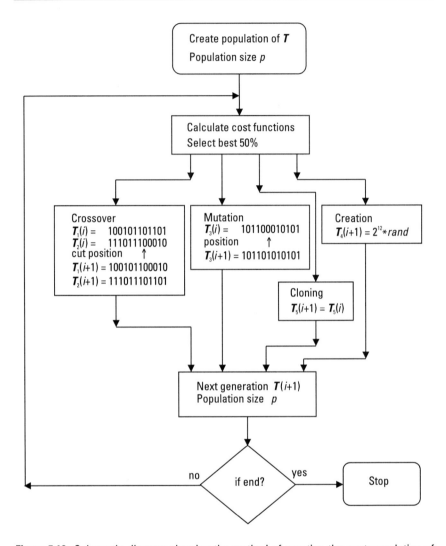

Figure 5.16 Schematic diagram showing the method of creating the next population of
$T(i + 1)$ from $T(i)$ in the genetic algorithm method of optimization. The four
processes of crossover, mutation, creation, and cloning are shown acting
on the binary representation of the variable parameters. The vector T is represented by a 12-bit binary number.

One important parameter in the genetic algorithm is the selection of
the population size. Goldberg et al. [13] calculated that an approximate value
for the desirable population size p is:

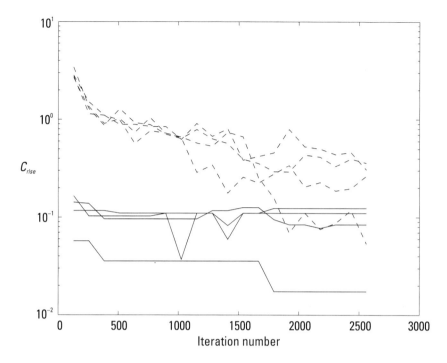

Figure 5.17 Variation in the mean and minimum values of C_{rlse} for four runs of a simple genetic algorithm optimization of the design task. The iteration number used on the x axis is the product of the population size p and the generation number. '____' minimum, '_ _ _' mean.

$$p \cong \frac{l}{k}\chi^{k} \qquad (5.23)$$

where l is the length of the chromosome, k is the average length of the parameters in the chromosome, and $\chi = 2$ for a binary representation of the chromosome. Others [14] suggested the range $30 < p < 100$ was appropriate for electromagnetic problems.

A number of different criteria can be used to terminate the optimization process. For example, the number of cycles might be fixed or the process terminated when the minimum value of the cost function in the newest population is not significantly different from that in the previous generations.

The GA technique was applied to the design task in Section 5.1.4. In (5.23), $l = 12$, $k = 2$, and so the population size $p = 128$. The subsequent generations were derived in the following way: (a) 50% of the old population, derived from crossover, using the best 50% of the previous generation, (b) 47% from cloning, and (c) 3% from creation.

The routine was run through 20 generations for four different runs, with a randomly selected starting population using (5.13). Figure 5.17 shows the change in the mean value and minimum value of the cost function C_{rlse} for the total population for four different runs of the algorithm. The iteration number (x axis) is the number of times the solver was used in the optimization process, and allows for a direct comparison with other optimization techniques.

While the mean value of the cost function decreases over the first 2 to 3 generations, the minimum value does not always show an improvement, even if the global minimum has not been located. The use of creation in each generation results in a significant and ongoing variation in the mean value of C_{rlse}, but often there is little change in the minimum value of C_{rlse} after the first few generations. There is a finite probability that the minimum value of C_{rlse} can increase, showing that the best T value has been lost during the selection of the next generation. This is always a possibility because there are very few points around the minimum value for this problem (see Figure 5.4). Figure 5.17 indicates that each run of the GA has resulted in a different final result, and that only one terminated in the general vicinity of the global cost function minimum value of $C_{rlse} = 0.017$.

For problems where the discretization of the variable parameters is relatively coarse in relation to variations in the cost function, the likelihood of locating and remaining in the global minimum may not be sufficiently high. There are two solutions to this problem. The first is to store all cost function data. At the end of the routine, this data set can be searched for the lowest value of the cost function. An alternative solution is to increase the resolution of the discretization by increasing the number of bits in the representation of the input parameters T. This is generally reliable, but requires additional computational effort. The level at which this discretization is applied requires an understanding of the slope of the cost function in the vicinity of the global minimum. This is never known with any certainty, particularly if the problem is a new one. This is a significant disadvantage of the GA optimization method. Another disadvantage is that the final result is not likely to be the optimal input parameter T, but rather, a point that lies close to the optimal position. This is the case with each run terminating at a different cost function value. It may be necessary to use an alternative technique to precisely locate the global minimum after the GA has been run.

5.7 Simulated Annealing

The simulated annealing optimization technique was developed at IBM to solve multiparameter optimization problems with sufficient robustness to avoid local minima [23–26]. It is an incremental-step-based technique, similar to the simplex method described in Section 5.4, but incorporates the possibility of stepping in the "wrong" direction. The technique is analogous to the annealing or tempering of steel, where the initial temperature of the steel is high and the temperature is reduced in a controlled manner to the point where the metal becomes completely rigid and its crystal structure is locked in place. In this optimization process, the assigned annealing "temperature" of the process equates to the probability of stepping from a lower cost function value to a higher one, that is, to progress in a direction away from the minimum value. As the temperature of the system is reduced, this probability declines, and so the number of transitions to higher values of the cost function decreases. In the limit this probability changes from 50% initially to 0% when the process terminates.

The basic technique shown in Figure 5.18 requires an initial value for T selected randomly using (5.13) and the calculation of the cost function for this vector C_{rlse}. The next value of T lies at an adjacent position one increment away. The direction of the step is selected randomly. Thus, if there are n variables in T, then there is a $(100/2n)\%$ chance of selecting a change in a specific direction. For example, if $n = 2$, the probability of the next value being at $\{t_1(i), t_2(i) + \Delta t_2\}$ is 25%. The new cost function at the point $(i, j + 1)$ is then determined. If $C_{rlse}(i, j + 1) < C_{rlse}(i, j)$, then $T(i + 1)$ is accepted as the new basis for comparison. If, however, $C_{rlse}(i, j + 1) > C_{rlse}(i, j)$, then there is a finite probability that this will be accepted as the new basis for comparison. The probability P that this is accepted, even though it has a higher cost function compared with the previous value, is given by the function:

$$P = e^{-d/T} \tag{5.24}$$

where T is the temperature of the solution, and $d = C_{rlse}(i + 1) - C_{rlse}(i)$ is the increase in the cost function resulting from the acceptance of the new value.

Equation (5.24) can be converted into a computer-generated random binary bit b that has probability of P of being unity, using the following expressions:

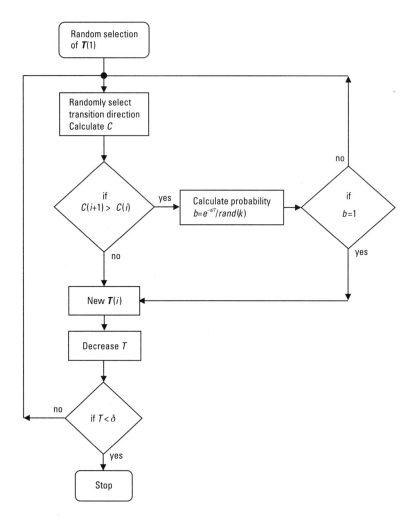

Figure 5.18 Iteration technique used to select the next **T** value using the simulated annealing technique.

$$\text{if } P \,/\, rand(k) \geq 1 \text{ then } b = 1$$
$$\text{if } P \,/\, rand(k) < 1 \text{ then } b = 0 \qquad (5.25)$$

where $rand(k)$ is a random number lying between zero and unity.

The temperature T of the system is decreased as the iteration number increases. The simplest method of decreasing T is an exponential cooling scheme [23] where:

$$T_{k+1} = \alpha T_k \qquad (5.26)$$

where α is a constant close to but smaller than unity. For example, Kirkpatrick et al. [23] used $\alpha = 0.95$. Another scheme uses a linear change [26] where:

$$T_{k+1} = T_k - \Delta T \qquad (5.27)$$

T should never be allowed to become negative, as then $P > 1$, meaning that the probability of a transition is greater than 1.

The possible options for when to change T include: (a) reducing T after a fixed number of iterations at one temperature state, (b) counting the number of positive cost-function transitions and reducing the temperature after a fixed number of such transitions, and (c) waiting for a specified decrease in the cost function. The routine is terminated when the temperature is so low that no further positive cost function transitions are possible and no steps can be taken that result in a decrease in the cost function.

The technique was applied to the design task described in Section 5.1.4 using (5.26) for the temperature change with $\alpha = 0.95$. The initial temperature chosen was $T = 0.001$, and this was changed after every 50 steps. The process was run four times using different start values. Figures 5.19 and 5.20 show the progression through the solution space and the variation in the cost function with iteration number, respectively. The process was terminated when no further step was possible.

Two of the four optimization attempts successfully located the cost function minimum. The other two runs terminated at local minimum values. Further adjustments in the initial temperature and the temperature change technique might ensure a more reliable outcome. The rapid variations of the cost function in Figure 5.20 are the result mainly of the unsuccessful search for a new step in one of the parameters. The step direction is random and consequently the cost function must be evaluated to decide whether this new step is acceptable. The apparently large variations in the cost function in Figure 5.20 show the sensitivity of the cost function to the finite step size used.

5.8 Summary

In this chapter, we explored six approaches to optimization. For illustration, the techniques were all applied to the relatively simple design task given in

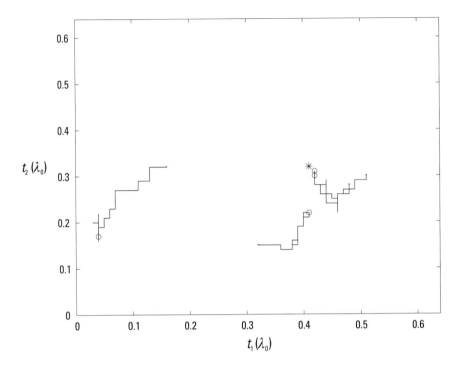

Figure 5.19 Passage of the updated parameter **T** through the solution space for the simulated annealing optimization technique. Four different runs with different start values are shown and their termination points are marked by o. The global minimum, at $t_1 = 0.42\lambda_0$ and $t_2 = 0.31\lambda_0$, is marked *.

Section 5.1.4. In this problem, two different spatial values in an antenna design were optimized by locating the minimum in a cost function. The cost function was created from the antenna front-to-back ratio, the impedance match, and the beamwidth. In all cases, the number of calls to the solver was logged and plotted. This allows a direct comparison between the techniques. It has been emphasized that all methods have been implemented in their simplest form and that further improvements are available and described in the literature.

Despite the simple antenna structure, the cost function is not monotonic across the solution space in the direction of the cost function minimum. A number of the techniques used here are prone to termination in local minimum values.

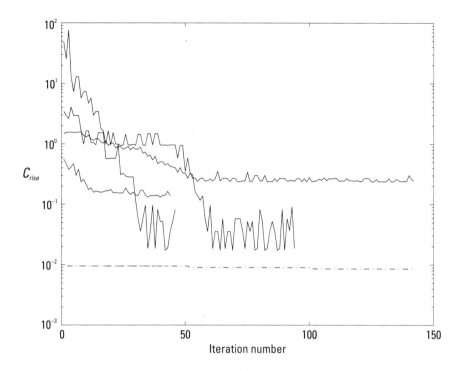

Figure 5.20 Variation in cost function as the simulated-annealing algorithm progresses. Four different start values were used.

For the guided optimization methods, the plots showing the variation in the cost function C_{rlse} versus the iteration number give an indication of the speed of convergence. This is one basis for comparison between the techniques. It must be emphasized, however, that the comparison presented here is for the solution of the problem given in Section 5.1.4 only. It is by no means an exhaustive set of tests of the optimization process.

As the cost function surface shown in Figures 5.3 and 5.4 is undulating, it is tempting to think that the optimization problem is almost trivial because the minimum value is clearly visible. This proved not to be the case, with a high probability that all guided algorithms failed to provide the correct, unique answer after one run of the procedure. It is evident that the "ridges" and "troughs" in the cost function surface result in the location of local minimum values rather than the relatively narrow global minimum. From the sequential uniform sampling technique plot of T versus the cost function (Figure 5.4), showing the complete sample space of 64×64 points, less than 20 points are located in the global minimum. A single point randomly selected (the Monte Carlo method) therefore has a probability of 20/4,096 in locating one of these points close to the minimum, and a 400-point Monte Carlo method has a relatively high probability of including one of these points.

It is evident from the cost function iteration data that most methods have a requirement that the cost function is evaluated at adjacent points to conclude that the minimum value has been reached.

The range of the cost function is not known for a new optimization problem. A small number of points selected randomly may give a single low value of the cost function, but this does not necessarily imply that this point is close to the global minimum.

Table 5.4 gives a semiquantitative comparison of the techniques based on the results presented in earlier sections.

Table 5.4 does not provide conclusive results about which is the best overall optimization technique. All but the sequential uniform sampling method rely on some aspect of randomization, and consequently the first values selected have a significant influence on the speed of convergence. For this reason, designers might include their "best guess" of the solution as the start position.

The design task used for these examples involved only two variable design parameters (see Figure 5.1), and therefore it is possible to represent the cost function surface graphically. For a greater number of variable design parameters, this visualization is not possible. Once the local minima have been avoided, the gradient method offers the most direct method of obtaining the global minimum, although the technique suffers from local minima trapping. The simulated annealing technique required the smallest number of calls to the solver, but also suffered from local minimum trapping. As the statistics of small numbers is prone to wrong conclusions, any attempt must be treated with extreme caution.

Despite this, the following conclusions can be made. For this problem, a number of runs of the simulated annealing algorithm proved the most effective technique in locating the cost function minimum. Without prior knowledge of the cost function distribution, it would not be unusual to run the technique with different start values many times to verify that the lowest cost function value obtained was the global minimum in the solution space. With sufficient computing power, the sequential linear sampling technique should never be overlooked, as it provides the most comprehensive data set.

For large numbers of parameters in T, the current opinion is that either the genetic algorithm or simulated annealing methods offer most promise.

Table 5.4
Comparison Between the Various Optimization Techniques Used to Solve the Design Task
Outlined in Section 5.1.4*

Method	Iteration Strategy	Randomization	Number of Iterations
Sequential Uniform Search	Unguided All points used	Not used	4,096
Monte Carlo	Unguided Random selection of points	Selection of all T	400
Simplex	Guided by reflection, expansion, and contraction	Selection of T for start value	1,000
Gradient	Guided by gradient	Selection of T for start value	120
Genetic Algorithm	Guided by allowing "best" of population to create children	Selection of population of T, then random selection for crossover and mutation	1,800
Simulated Annealing	Guided by probabilistic determination of direction of next step with decreasing temperature values	Selection of T for start value, and random selection of direction of each new point	35

* The number of iterations refers to the number of times the solver was used to calculate the cost function.

While the GA technique is more computationally intensive than simulated annealing, the number of variable design parameters in T can be readily increased, and the resolution of those parameters can also be increased without a significant change in the computational effort required. There is no clear superior method.

All optimization problems are ideally suited to parallel processing. The techniques benefit from the use of more than one computer running in parallel, for example, through the use of a different start position. In particular, methods of avoiding local minima require strategies in which a random factor is introduced, either through the use of a randomly selected start location T, or by using a technique that relies on some randomization within the process itself. In this way, local minima may be avoided. Inherent in every random process is the possibility that the global minimum in the cost function will be missed. Users of any optimization technique (apart from the sequential uniform sampling technique) will find it beneficial to run the routine more than once, if not a large number of times.

References

[1] Numerical Electromagnetics Code (NEC). This code was originally developed at the Lawrence Livermore National Laboratories, California, and is now available from R. Anderson, 2000, http://www.qsi.net/wb6tpu/suindex.html.

[2] Ansoft Corporation, Ensemble 6.1, Pittsburg, PA, 2000, http://www.ansoft.com.

[3] Diaz, L., and T. Milligan, *Antenna Engineering Using Physical Optics, Practical CAD Techniques and Software*, Norwood, MA: Artech House, 1996.

[4] Taflove, A., *Computational Electrodynamics: The Finite-Difference Time-Domain Method*, Norwood, MA: Artech House, 1995.

[5] Kunz, K. S., and R. J. Luebbers, *The Finite Difference Time Domain Method for Electromagnetics*, Boca Raton, FL: CRC Press, 1993.

[6] Ansoft Corporation, HFSS, Pittsburg, PA, 2000, http://www.ansoft.com.

[7] Nittany Scientific, Nec Win Pro 1.1, Riverton, UT, 1999, http://www.nittany-scientific.com/nwp/index.htm.

[8] Sobol, I. M., *A Primer for the Monte Carlo Method*, Boca Raton, FL: CRC Press, 1994.

[9] Daniels, R. W., *An Introduction to Numerical Methods and Optimization Techniques*, New York: North Holland, 1978.

[10] Spendley, W., G. R. Hext, and F. R. Humsworth, "Sequential Applications of Simplex Designs in Optimization and Evolutionary Operation," *Technometrics*, Vol. 4, 1962, pp. 441–461.

[11] Everitt, B. S., *Introduction to Optimization Methods and Their Application in Statistics*, London: Chapman Hall, 1987.

[12] Nelder, J. A., and R. Mead, "A Simplex Method for Function Minimization," *Computer Journal*, Vol. 7, 1965, pp. 308–313.

[13] Goldberg, D. E., K. Deb, and J. H. Clark, "Genetic Algorithms, Noise, and the Sizing of Populations," *Complex Systems*, Vol. 6, 1992, pp. 333–362.

[14] Johnson, J. M., and Y. Rahmat-Samii, "Genetic Algorithms in Engineering Electromagnetics," *IEEE Antennas and Propagation Magazine,* Vol. 39, No. 4, August 1997, pp. 7–21.

[15] Michielssen, E., Y. Rahmat-Samii, and D. S. Weile, "Electromagnetic System Design Using Genetic Algorithms," in *Modern Radio Science,* M. A. Stuchly (ed.), New York: Oxford University Press, 1999.

[16] Weile, D. S., and E. Michielsson, "Genetic Algorithm Optimization Applied to Electromagnetics: A Review," *IEEE Trans. Antennas and Propagation,* Vol. 45, No. 3, 1997, pp. 343–353.

[17] Haupt, R. L., "An Introduction to Genetic Algorithms for Electromagnetics," *IEEE Antennas and Propagation Magazine,* Vol. 37, No. 2, 1995, pp. 7–14.

[18] Yegin, K., and A. Q. Martin, "Very Broadband Loaded Monopole Antennas," *IEEE Antennas Propag. Intl. Symp.,* Montreal, Cananda, 1997, pp. 232–235.

[19] Yan, K. K., and Y. Lu, "Sidelobe Reduction in Array-Pattern Synthesis Using Genetic Algorithm," *IEEE Trans. Antennas and Propagation,* Vol. 45, No. 7, 1997, pp. 1117–1122.

[20] Naval Research Laboratory, *Practical Guide to Genetic Algorithms,* 2001, http://chemdiv-www.nrl.navy.mil/6110/sensors/chemometrics/gademo.html.

[21] Goldberg, D. E., *Genetic Algorithms in Search, Optimization and Machine Learning,* Reading, MA: Addison Wesley, 1989.

[22] Darwin, C., *The Origin of Species,* London: Penguin Books, 1985.

[23] Kirkpatrick, S., C. D. Gelatt, and M. P. Vecchi, "Optimization by Simulated Annealing," *Science,* Vol. 220, May 1983, pp. 671–680.

[24] Otten, R. H. J. M., and L. P. P. P. van Ginneken, *The Annealing Algorithm,* Norwell, MA: Kluwer Academic Publishers, 1989.

[25] http://csep1.phy.ornl.gov/csep/MO/NODE28.html.

[26] Randelamn, R. E., and G. S. Grest, "N-City Travelling Salesman Problem—Optimization by Simulated Annealings," *J. Statistical Physics,* Vol. 45, 1986, pp. 885–890.

6

Performance Limitations

6.1 Introduction

In Chapter 2, we saw that the beamwidth of an antenna was related to its physical size. This fundamental limit to the beamwidth was discussed in relation to the overall size of an antenna.

In earlier chapters, the antenna performance was usually calculated assuming that the materials and components used in their construction had ideal characteristics. That is, the conductivity of a conductor was assumed to be infinite, and no loss term was included in the dielectric constant. The performance of the switches in the switched parasitic antennas was also considered to be ideal, that is, in the open-circuit position the capacitance was zero, and in the short-circuit position, the resistance was zero. These assumptions are not always valid. In this chapter, the effects of nonideal performance are considered. The opportunities offered by a number of new technologies are also considered as methods for overcoming these problems in switched parasitic antennas.

6.2 Beamwidth Limits

In Chapter 5, two groups of input design parameters were distinguished; the fixed design parameters established the basic structure of the antenna, and the variable design parameters were adjusted in the optimization process.

The use of optimization techniques to improve the performance of a basic antenna design can only accomplish improvements within certain absolute constraints. For example, the radiation characteristics are constrained by the mechanical dimensions of the antenna.

General limits to the radiation pattern can be derived from a knowledge of the physical size of an antenna. They can be investigated through Fourier transform analysis of the current distribution in the antenna volume as outlined in Section 2.3. A first approximation to the E plane radiation pattern is obtained from (2.17), the Fourier transform of the spatial current distribution. It is not necessary to ensure that the current distribution approaches zero at the edges of the structure in the Fourier transform plane, because a three-dimensional object allows the current to flow along a conductor normal to this plane. An example of this is a rectangular loop antenna, where the two sides in parallel are the only currents in the plane of interest. The $\sin\theta$ term that lies outside the integral in (2.17) remains important, particularly in the case of electrically small antenna structures.

Assuming that the current is uniform across the antenna, then the E plane radiation pattern is given by the sinc function in (2.24). The number of lobes in the radiation pattern can be calculated from the number of zeros in the sinc function. This is determined by kl, the product of the wave number k and the dimension of the length of the antenna l. The number of zeros in the radiation pattern m lying in the angular range $0 < \theta < \pi/2$ can be written as:

$$m = Int\left(\frac{l\cos\theta}{\lambda_0}\right) \qquad (6.1)$$

where $Int(a)$ is the integer value of the expression a rounded down.

If $l < \lambda_0$, and $\cos\theta < 1$, the maximum number of radiation minima in the pattern is $m = 1$. There are two additional minimum values at $\theta = 0$ and $\theta = \pi$. This has implications for the beamwidth of the antenna, although the position of the minimum may lie close to either $\theta = 0$ or $\theta = \pi$.

For antennas with dimensions $l < \lambda_0$, the position of the minimum value in the radiation pattern is dependent on the current distribution in the antenna and its overall length l. It is possible to achieve 10-dB beamwidths in the range $180° < B_\theta < 270°$. For smaller beamwidths and consequently higher gain, the size of the antenna must be increased and the current distribution optimized appropriately.

For a larger antenna with $l > \lambda_0$, the number of minimum values in the radiation pattern increases, but m remains dependent on the current distribution.

In the H plane radiation pattern, for a single wire antenna, the current is described by a delta function, (2.10), and the H plane radiation pattern is uniform across all ϕ angles. If there are two wire antennas in the array and the distance between the wires is less than λ_0, then the number of minimum values in the H plane radiation pattern is $m = 1$.

These general principles are true for all antenna systems, including parasitic antennas, phased arrays, and aperture antennas. When discussing directional antennas, the beamwidth is usually defined in terms of the -3-dB or -10-dB radiation pattern characteristics rather than the angular distance between the minimum values in the radiation patterns [e.g., (2.1)]. The signal strength in the minor beams plays a significant role in the overall performance of the antenna, and so it may be necessary to use these values in the cost function when using the optimization routines described in Chapter 5.

6.3 Antenna Losses

6.3.1 Conductor Losses

In this book it has been assumed that the conducting elements and ground planes are lossless. This implies that the conductivity σ of all conductors in the antenna is infinite. This is not the case in practice, as the metals commonly used in antenna construction have a conductivity $\sigma < 10^8$ S/m. Some coatings used to protect the metal antenna from the environment (e.g., zinc in the galvanizing process) can have much lower conductivities.

At high frequencies, the RF current is confined to the outer surface of the conductor and so the properties of the coating material dominate the conduction process. The depth of penetration of the current beneath the outer surface of the conductor is referred to as the skin depth δ and can be calculated from the equation:

$$\delta = \sqrt{\frac{1}{\pi f \mu \sigma}} \tag{6.2}$$

where μ is the magnetic permeability of the conductor and f is the frequency. The conductivities of a number of metals at 20°C are listed in Table 6.1.

Table 6.1
Conductivity Values of Various Metals Used in Antenna Construction

Metal	Conductivity (S/m)
Gold	4.25×10^7 [1]
Copper	5.85×10^7 [1]
Aluminum	3.50×10^7 [1]
Steel	$0.14 \times 10^7 - 0.96 \times 10^7$ [2]
Stainless steel	$0.09 \times 10^7 - 0.11 \times 10^7$ [3]
Silver	6.30×10^7 [1]
Zinc	1.69×10^7 [2]
Tin	0.88×10^7 [1]

The finite conductivity of the metals used in antenna construction contributes to an additional resistive component in the input impedance of the antenna. The effect of these loss terms is usually to increase the bandwidth of the antenna and decrease the minimum value of S_{11} in dB. The variation in S_{11} for a center-fed wire dipole is shown in Figure 6.1 for a number of different conductivity values. While a wire conductivity of 10^8 S/m has characteristics close to the perfect case (i.e., infinite conductivity), there is significant degradation as the conductivity is decreased beyond this value.

6.3.2 Dielectric Losses

All dielectric materials have a finite loss term. The loss angle δ_c and the loss tangent $\tan\delta_c$ are defined in terms of the ratio of the conduction current term to the displacement current term in the expression for complex permittivity [1, 2], that is,

$$\tan \delta_c = \sigma \,/\, 2\pi f \varepsilon_r \varepsilon_0 \tag{6.3}$$

where σ is the conductivity of the material derived from all loss mechanisms, ε_r is the relative permittivity of the material, ε_0 is the permittivity of free space, and f is the frequency of the radiation.

This loss occurs in the insulating materials used to support the conducting elements and the dielectric material used in the construction of

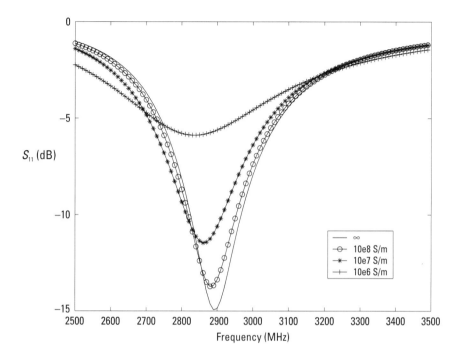

Figure 6.1 Variation in S_{11} (dB) for a center-fed dipole of length 0.05m and radius 0.0001m for various values of the conductivity of the wire.

patch antennas. The relative permittivity values for some typical materials used in antenna construction are listed in Table 6.2 [2–5]. There is also a very wide range of dielectric substrates produced by a number of companies for use in microwave circuit construction and microwave patch antennas [1].

The use of a lossy dielectric material reduces the S_{11} (dB) value of a patch antenna at resonance. This is illustrated in Figure 6.2 for a 55 mm × 55 mm probe-fed patch on a 1.6-mm-thick lossy substrate calculated using Ensemble [6]. The gain of the patch antenna is also reduced substantially, from a gain of 7.3 dBi when $\tan\delta_c$ is 0 to 4.8 dBi when $\tan\delta_c$ is 0.01.

The bandwidth of an antenna can be increased by using lossy materials in the structure [7]; however, this is accompanied by a decrease in the radiated power. Alternatively, the bandwidth of an antenna is reduced when materials with a high permittivity and very low loss are placed close to the conducting elements. This can be a problem with some dielectric resonator antennas.

Table 6.2

Relative Permittivity ε_r and Loss Tangent $\tan\delta_c$ for Dielectric Materials Used
in Antenna Construction

Material	Relative Permittivity ε_r	Loss Tangent $\tan\delta_c$ @ 1 GHz
Glass	3.9–9.5 [2]	0.0007–0.04 [2]
PVC	3.3 (1 MHz) [2]	0.0055 (3 GHz) [3]
	2.84 (3 GHz) [3]	
Nylon	2.84 (3 GHz) [3]	0.117 (3 GHz) [3]
	3.4 (100 MHz) [2]	
Air	1.0006 [4]	
Teflon	2.1 (1 kHz–3 GHz) [5]	1.5×10^{-4} (3 GHz) [5]
Alumina	4.5–8.4 [2]	0.0002–0.01 [2]
Porcelain	6.0–8.0 [2]	0.006–0.02 [2]
Rubber	2.15 –4 (3 GHz) [3]	0.0009–0.0097 (3 GHz) [3]
	2.4–4.5 (100 MHz) [2]	
Plexiglass	2.6 (3 GHz) [5]	57×10^{-4} (3 GHz) [5]

6.3.3   Other Losses

Additional losses occur in antenna systems where passive RF circuit elements are used. For example, in Section 2.9, lumped *R, L,* and *C* elements were used to increase the number of discrete frequency bands. Each of these elements introduces loss terms that influence the overall performance of the antenna. Other passive elements such as transmission line cable connectors, RF switches, *p.i.n.* diodes, rotary joints, power splitters, and power combiners all have a nonzero insertion loss that reduce the power radiated by the system. For example, at 10 GHz, commonly used coaxial connectors of type SMA have an insertion loss of 0.1 dB; for type N the insertion loss is 0.16 dB [1]. The precision design and manufacture of these components is such that the cost of these items is often very high. It is often necessary to reduce the number of passive RF components in the total antenna system, and to consider integrated manufacturing techniques such as microstrip circuits rather than an assembly of coaxial components [1].

In the case of dual-band antennas using *RLC* elements, the performance of the discrete components is often frequency dependent. The impedance value of an RF inductor can change by more than 30% over the frequency range 900 MHz to 1,200 MHz [8]. These variations must be considered as part of the design process.

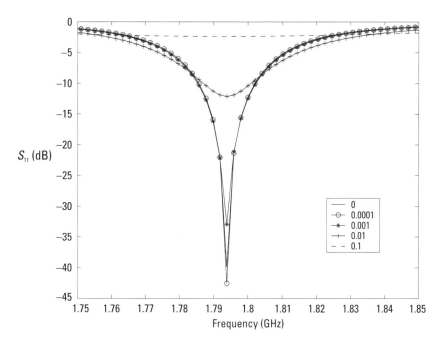

Figure 6.2 Variation in S_{11} (dB) for a probe-fed patch antenna (55 mm x 55 mm) on a 1.6-mm-thick substrate with relative permittivity ε_r = 2.2 and various values of the loss tangent.

6.4 Switching Circuits for Switched Parasitic Antennas

6.4.1 FASPA Switches

RF switches such as *p.i.n.* diodes are an important building block for switched parasitic antennas. There are two different switching requirements. FASPAs require single-pole, single-throw (SPST) switches to change the parasitic element from a resonant condition to a nonresonant condition and back again. These switches have two ports. In modeling these antennas, it has been assumed that the performance of the switch is ideal. This means assuming that when the switch is closed, the impedance is zero, and when the switch is open, the impedance is infinite. Because the signals are of a very high frequency, the latter assumption implies that the capacitance for the open switch is zero. The gap between the two parts of the antenna element must be large for this to be true. There will always be a parasitic capacitance between the two sections; however, this capacitance will be very small if the radius of the wire is sufficiently small and the gap is sufficiently large.

Consider two parallel plates of area A separated by a distance d with a dielectric material between the plates having a relative permittivity ε_r. If the fringing fields are ignored, the capacitance C between the plates can be written [5]:

$$C = \varepsilon_r \varepsilon_0 A / d \qquad (6.4)$$

For example, in a parasitic array with an open-circuited feed position and no diodes, if the wire has a radius of 5 mm, and the air gap is 1 mm, the capacitance is approximately 0.7 pF. The reactance of this capacitance at 3 GHz is 480Ω. This value is at the center of the transition zone for a lumped-impedance element that is used to change the effective length of a wire antenna (see Figure 2.29). This would constitute a poor-quality open circuit in the parasitic element and degrade the performance of the switched parasitic antenna.

For a reverse-biased semiconductor diode junction, (6.4) can be used to calculate the capacitance across the diode. In this case d is the width of the depletion region between the p-type and n-type semiconductor material in the diode [9, 10]. The value of d is dependent on the applied reverse-biased voltage V and the carrier density of the two doped materials. The relationship can be written:

$$C = kV^{-1/2} \qquad (6.5)$$

and k is a constant calculated from the doping charge densities in the semiconductor junction.

The distance d in a semiconductor junction can be increased by including a layer of undoped (or weakly doped) semiconductor material called intrinsic material, to lie in between the p and n type material in a diode. This forms a $p.i.n.$ diode and has the property of a much lower reverse-biased capacitance value when compared with a standard p-n junction [10]. It has the additional, undesirable effect of increasing the resistance of the forward-biased diode.

The equivalent circuit of a $p.i.n.$ diode is given in Figure 6.3 [10], where C_d and C_b are stray capacitances from the circuit parasitic capacitance and the case capacitance respectively. L_s is the inductive component from the connecting conductors in the package. The forward bias resistance R_{SF} is dependent on the applied bias voltage. The selection of a diode for RF switching is a compromise. For a series-connected diode switch, it is desirable to

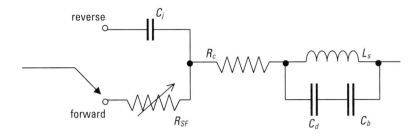

Figure 6.3 Equivalent circuit of a *p.i.n.* diode for both forward- and reverse-biased operating conditions.

minimize both the forward-biased resistance ($R_c + R_{SF}$) and the reverse-biased capacitance C_j. The cross-sectional area of the semiconductor junction A is directly proportional to C_j, and is also inversely proportional to the contact resistance R_c.

Diodes fabricated from gallium arsenide can have switching speeds of up to 2 nS with a forward resistance of 2Ω and a maximum capacitance of less than 0.1 pF [10–12]. With these characteristics, operation at 10 GHz is possible and the device can be driven from TTL logic circuits. The switching speed is of the same order as the settling time of typical diode detector circuits, and so is not a major limitation to the overall scan speed of the antenna. This issue was discussed in Chapter 1.

The switching circuits for a switched parasitic monopole antenna element must be designed so that the dc voltage connections do not influence the radiation or impedance characteristics of the antenna. In addition, the current through the diode must be limited to prevent overvoltage-induced breakdown, and in portable applications, to minimize power consumption. For these reasons, resistive lead wires (or lumped resistive elements) and RF choke elements are used in series with the switching diode. A typical circuit for a switched parasitic element driven by a TTL voltage from a digital logic control circuit is shown in Figure 6.4 [13].

A switched parasitic antenna in transmit mode may be required to handle significant RF power levels. In this case, the currents in the shorted parasitic elements can be high and the voltage across the open-circuit switches must be considered to ensure that breakdown does not occur. This will require the diodes to be reverse-biased so that the induced voltage across the terminal of the open-circuit elements does not turn the diode on. The overall current rating of the switch must be compared with the calculated current in the parasitic element. As these switches are not mounted directly across the

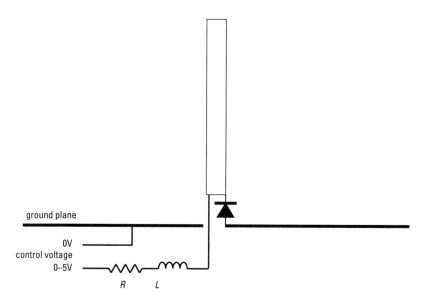

Figure 6.4 TTL-driven control circuit for a switched parasitic monopole. *R* limits the current through the diode and *L* prevents the RF signal in the antenna element from leaking below the ground plane.

driven element in a FASPA, this current requirement is much less than the current in the feed element.

6.4.2   SASPA Switches

SPST switches are required in SASPA antennas. In this case, the RF power is applied directly across the open circuit switch. The reverse-bias power rating of the switch must be adequate to ensure RF currents do not flow through the switch. In the case of a reverse-biased semiconductor switch, the voltage across the switch from the RF signal must be smaller than the bias voltage required to obtain an open circuit. This may require a significant reverse voltage across the diode.

SASPA antennas require single-pole, *n*-throw (SP*n*T) switches where *n* is the number of possible active elements. If there are four RF feed positions in a single beam antenna, there are four switch positions requiring both a SP*n*T in parallel with a SPST switch.

For transmitter applications, high-power switches are required. This restricts the selection of a *p.i.n.* diode switch to parallel diode circuits [10].

Integrated circuit switches with up to six throw positions (i.e., SP6T) fabricated in gallium arsenide technology are commercially available from a number of suppliers [11, 12] and have an integrated CMOS decoder and driver circuit suitable for TTL digital logic. This means that the switch can be controlled directly from a microcontroller circuit. Typical microwave switches rated at 2–3 GHz have a maximum insertion loss of less than 3 dB and isolation of approximately 30 dB. The power-handling capacity is approximately 30 dBm, with typical switching speeds of 3 ns. These characteristics make them suitable for operation in SASPAs.

For SASPA antennas, the phase of the signal delivered to the driven element is not important, and so complex phase-shifting systems are not required. In phased arrays, however, the situation is very different, and phase-shift networks are costly in terms of space, insertion loss, and the controller complexity required to align the array.

6.5 New Technologies and New Applications

6.5.1 MEMs RF Switches

While the performance of the semiconductor RF switches appears adequate in the current communications frequency bands, for frequencies greater than approximately 10 GHz, the impedance change between a short and open circuit is not adequate. The use of inductive relays is unsuitable at RF because the switching speed is slow and the activation current requirements can be large. In addition, the insertion loss of these circuit components is very poor. Recently there has been considerable activity in the design and fabrication of microelectromechanical (MEMs) RF switches [14, 15]. This technology relies on the fabrication techniques previously reserved for the semiconductor electronics industry. MEMs devices fabricated on a bulk semiconductor or insulating substrate using photolithography offer considerable promise in achieving ideal switch performance; that is, low short-circuit insertion loss and high open-circuit isolation. In this case, the switches are mechanical in operation and are based on the principle of a variable beam. Two basic structures have been used to date; a cantilever beam activated by an electrostatic charge, and an air bridge activated either by a heating element or an electrostatic charge [14]. These two structures are shown in Figure 6.5.

The very small size on the conduction lines in the switch and the relatively large air gap results in off-capacitances in the range 30–40 fF [15]. The on-resistance values are less than 1Ω. This is much better than any electronic switch and enables operation at frequencies up to 40 GHz [15].

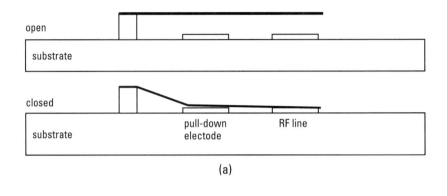

(a)

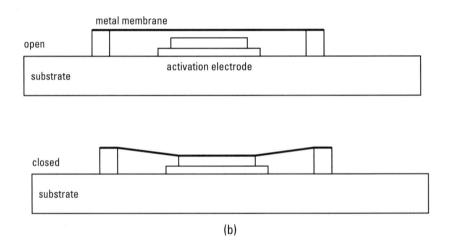

(b)

Figure 6.5 Schematic diagram of two styles of micromachined RF switches: (a) cantilever and (b) air bridge. The switches are both activated by electrostatic charge.

There are, however, technological problems with the use of MEMs switches that have yet to be fully overcome. In comparison to electronic switches, the switching speed is slow. This is principally a result of the mechanical action and the driving technique. For example, one heat-activated MEMs switch has a maximum switching frequency of 30 Hz, with a nominal control power of 180 mW [16]. An electrostatically activated switch is limited by the RC time constant of the circuit to charge and discharge the capacitor. The switches are still unreliable because of stiction—an effect where the switch becomes permanently closed through normal operation, often caused by humidity. In addition, there is usually a significant current drain from the drive circuit when the switch is closed. This remains a very active field of investigation, and there is certainly much optimism that these problems can be overcome in the near future.

6.5.2 Superconductor Technology

A second development likely to have a very positive influence on antennas is the advent of high-temperature superconductors. A variety of ceramic materials exhibit superconducting properties at temperatures as high as 90K. While the term "superconductor" implies that these materials are excellent conductors of electricity, they are more appropriately described in electromagnetic terms as materials with a large negative relative permittivity [17]. The permittivity of these materials is both frequency and temperature dependent. The real and imaginary parts of the relative permittivity, ε' and ε'', respectively, are found in the range $-10^{12} < \varepsilon' < -10^{9}$, and $1 < \varepsilon'' < 10^{7}$ [17].

The possible applications of this technology to antennas were reviewed in 1991 [18, 19]. Collins [18] concluded that the useful range of electrically small antennas could be improved through the use of superconductors in impedance-matching circuits, but there was no benefit in the antenna elements themselves being constructed of superconducting material. Dinger et al. [19] also concluded that the main application of superconductivity was in the significant reduction in conduction loss of impedance-matching components for antennas.

A number of antenna designs have been developed in which the superconducting material has been used in the antenna elements themselves, particularly through the use of meander-line patch antennas [20, 21]. As the size of conducting tracks becomes smaller, so then does the resistance of these tracks. The use of superconducting materials results in a reduced bandwidth and a higher resonant frequency when compared with similar structures fabricated using copper [18]. The bandwidth can be increased through the use of parasitic elements with slightly different resonant frequencies.

Since 1991, attention has also been directed at the problems of fabricating these ceramic materials in the required shapes and the encapsulation of the tuning circuits so that the elements remain at temperatures below the critical temperature of the materials. There is also ongoing research into the search for superconducting materials with even higher critical temperatures.

6.5.3   Smart Materials and MEMs Antennas

Smart materials are composite materials whose shape and rigidity are controlled by external stimulus. Researchers are investigating materials that respond to a wide variety of external forces such as electrical, thermal, chemical, pressure change, and vibration [22]. Communications systems have recently developed in frequency ranges above 20 GHz. The physical size of the antennas at these frequencies is such that the integrated fabrication of both antennas and active circuit elements on a single substrate has become an active area of investigation. The new class of antennas, called active arrays, has principally been directed toward the development of integrated electronics and antenna elements as part of a smart antenna. The use of smart materials and MEMs technologies to physically reconfigure the conducting elements in an array has also received attention [23].

One example of this technique [24] employs the use of piezoelectric drive actuators to vary the spacing between a microstrip patch and a parasitic patch located above it in a stacked patch configuration. This allows an electronic controller to change the bandwidth of an antenna operating at 3 GHz. The bandwidth was increased by a factor of 2 for a displacement of 5 mm. At higher frequencies the displacement requirements would be much smaller.

Micromachining allows the fabrication of a large number of passive elements on the same substrate. Single and stacked cavity-backed patch antennas operating at 20 GHz and 94 GHz, respectively, have been demonstrated with good radiation efficiency [25, 26].

The reduction in surface-wave interference in arrays of patch antennas has been achieved by modifying the effective dielectric constant of the substrate material. This allows thicker substrate materials with higher dielectric constant to be used, resulting in higher bandwidth and with reduced mutual coupling between antenna elements [27]. Based on the principle of photonic band-gap structures in which regular arrays of conducting and dielectric materials are used to control the bulk electromagnetic properties of the materials [28, 29], the material can be constructed in such a way that propagation through it is practically impossible. The net result is an increase in the bandwidth of the antenna as well as a decrease in the overall size of the array.

6.6   Conclusions

Switched parasitic antennas have many desirable attributes for smart antennas. These include small size, low power, and low cost. In this chapter, the limits to these attributes have been addressed. The size of the antenna does directly influence the beamwidth, and the radiation performance is limited by finite conductivity of the metals and dielectric loss.

As the frequency of communications systems increases with the corresponding decrease in wavelength, a variety of novel antenna classes based on new technologies will emerge. Switched parasitic antenna systems using MEMs switches are one class of antennas that offer significant promise to new technologies. The application of a combination of these new technologies might increase the complexity of the design process, but can lead to more efficient smart antenna systems.

References

[1] Laverghetta, T., *Microwave Materials and Fabrication Techniques*, 2nd ed., Norwood, MA: Artech House, 1991.

[2] *Handbook of Chemistry and Physics*, 56th ed., Cleveland, OH: CRC Press, 1975.

[3] Jordan, E. C., *Reference Data for Engineers: Radio, Electronics, Computer, and Communications,* Carmel, IN: Sams Publishing, 1985.

[4] Giger, A. J., *Low-Angle Microwave Propagation: Physics and Modeling*, Norwood, MA: Artech House, 1991.

[5] Paul, C. R., K. W. Whites, and S. A. Nasar, *Introduction to Electromagnetic Fields*, 3rd ed., Boston: McGraw-Hill, 1998.

[6] Ansoft Corporation, Ensemble, 2000, http://www.ansoft.com.

[7] Balanis, C. A., *Antenna Theory: Analysis and Design*, 2nd ed., New York: John Wiley and Sons, 1997.

[8] Coilcraft Inc., 2000, http://www.coilcraft.com.

[9] Havill, R. L., and A. K. Walton, *Elements of Electronics for Physical Scientists*, 2nd ed., London: MacMillan, 1985.

[10] Straeli, C., J. V. Bouvet, and D. Goral, "P-i-n and Varactor Diodes," in *The Microwave Engineering Handbook*, Vol. 1, B. L. Smith and M. H. Carpenter (eds.), New York: Van Nostrand Reinhold, 1993.

[11] M/A-COM Inc., *RF and Microwave Semiconductors,* Lowell, MA, 1995.

[12] Minicircuits Inc., 2000, http://wwww.minicircuits.com.

[13] Thiel, D. V., S. G. O'Keefe, and J. W. Lu, "Electronic Beam Switching in Wire and Patch Antenna Systems Using Switched Parasitic Elements," *IEEE Antennas and Propagation Symposium*, Baltimore, MD, 1996, pp. 534–537.

[14] Brown, E. R., "RF-MEMS Switches for Reconfigurable Integrated Circuits," *IEEE Trans. Microwave Theory and Techniques*, Vol. 46, November 1998, pp. 1868–1880.

[15] Goldsmith, C. L., Z. Yao, S. Eshelman, and D. Denniston, "Performance of Low-Loss RF MEMs Capacitive Switches," *IEEE Microwave and Guided Wave Letters*, Vol. 22, No. 8, August 1998, pp. 269–271.

[16] Cronos Integrated Microsystems Inc., "Ten-Switch Microarray CRL-10A5D24-6.0," 1999, http://www.memsrus.com.

[17] Mei, K. K., and G. Liang, "Electromagnetics of Superconductors," *IEEE Trans. Microwave Theory and Techniques,* Vol. 39, September 1991, pp. 1545–1552.

[18] Hansen, R. C., "Antenna Applications of Superconductors," *IEEE Trans. Microwave Theory and Techniques*, Vol. 39, September 1991, pp. 1508–1512.

[19] Dinger, R. J., D. R. Bowling, and A. M. Martin, "A Survey of Possible Passive Antenna Applications of High-Temperature Superconductors," *IEEE Trans. Microwave Theory and Techniques,* Vol. 39, September 1991, pp. 1498–1507.

[20] Lancaster, M. J., *Passive Microwave Device Applications of High Temperature Superconductors*, Cambridge, U.K.: Cambridge University Press, 1997.

[21] Wang, H. Y., and M. J. Lancaster, "Aperture-Coupled Thin-Film Superconducting Meander Antennas," *IEEE Trans. Antennas and Propagation*, Vol. 47, May 1999, pp. 826–836.

[22] Rogers, C. A., "Intelligent Materials," *Scientific American,* September 1995, pp. 154–155.

[23] Chiao, J. C., "MEMs RF Devices for Antenna Applications," *Proceedings of Asia Pacific Microwave Conf.*, Sydney, Australia, December 2000, pp. 895–898.

[24] Keily, E., G. Washington, and J. Bernhard, "Design and Development of Smart Microstrip Patch Antennas," *Smart Materials and Structures,* Vol. 7, December 1998, pp. 792–800.

[25] Papapolymerou, I., R. F. Drayton, and P. B. Katehi, "Micromachined Patch Antennas," *IEEE Trans. Antennas and Propagation*, Vol. 46, February 1998, pp. 275–283.

[26] Gauthier, G. P., et al., "A 94 GHz Aperture-Coupled Micromachined Microstrip Antenna," *IEEE Trans. Antennas and Propagation*, Vol. 47, December 1999, pp. 1761–1766.

[27] Colburn, J. S., and Y. Rahmat-Samii, "Patch Antennas on Externally Perforated High Dielectric Constant Substrates," *IEEE Trans. Antennas and Propagation*, Vol. 47, December 1999, pp. 1785–1794.

[28] King, R. J., D. V. Thiel, and K. S. Park, "The Synthesis of Surface Resistance Using an Artificial Dielectric," *IEEE Trans. Antennas and Propagation,* Vol. 31, March 1983, pp. 471–476.

[29] Sievenpiper, D., and E. Yablonovitch, "High-Impedance Electromagnetic Surfaces," in *Modern Radio Science*, M. A. Stuchly (ed.), New York: OUP, 1999, pp. 151–169.

About the Authors

David V. Thiel received a B.Sc degree in physics and applied mathematics from the University of Adelaide in South Australia. He subsequently received an M.Sc and a Ph.D. from James Cook University in Townsville, Queensland, Australia, in antenna design and electromagnetic geophysics, respectively.

Dr. Thiel moved to Griffith University in Brisbane, Australia, in 1974, where he set up the Radio Science Laboratory in 1983. As a result of a number of promotions, he is now a tenured professor in the School of Microelectronic Engineering, the director of the Radio Science Laboratory, and a program manager in the Cooperative Research Centre for Microtechnology. Dr. Thiel also holds two U.S. patents in the communications area: one concerned with modulation techniques and the other in switched parasitic antennas for cellular telephone handsets.

He has published more than 55 papers in the refereed scientific literature and presented more than 65 papers at international conferences. Dr. Thiel has served as an ADCOM member in the IEEE Antennas and Propagation Society and has been a member of the Wave Propagation Standards Committee since 1996.

His main research areas include switched parasitic antennas, computational techniques in electromagnetics, electromagnetic geophysics, smart environmental sensors using MEMs technology, robotics, and electronic odor sensing.

His teaching areas include introductory and advanced courses in electromagnetics, numerical techniques in electromagnetics, advanced communications systems, optical communications systems, VLSI design, microprocessor systems, and electronic systems.

Stephanie Smith received a B.Eng (honors) from Griffith University, Queensland, Australia, in 1996 and a Ph.D. from the same university in 2001. Her Ph.D. research area was electronically steerable antennas for mobile communications. Dr. Smith worked as a microwave design engineer for Mitec Pty. Ltd., Queensland, in 1996. She has been employed at CSIRO Telecommunications and Industrial Physics, Sydney, Australia, since December 1999; there she is working in the area of antenna and passive microwave circuit design. Dr. Smith is a member of the IEEE Antennas and Propagation Society and has published three journal papers and five conference papers.

Index

Recent Titles in the Artech House Antennas and Propagation Library

Thomas Milligan, Series Editor

Fresnel Zones in Wireless Links, Zone Plate Lenses and Antennas, Hristo D. Hristov

Handbook of Antennas for EMC, Thereza MacNamara

Iterative and Self-Adaptive Finite-Elements in Electromagnetic Modeling, Magdalena Salazar-Palma, et al.

Measurement of Mobile Antenna Systems, Hiroyuki Arai

Microstrip Antenna Design Handbook, Ramesh Garg, et al.

Mobile Antenna Systems Handbook, Second Edition, K. Fujimoto and J. R. James, editors

Quick Finite Elements for Electromagnetic Waves, Giuseppe Pelosi, Roberto Coccioli, and Stefano Selleri

Radiowave Propagation and Antennas for Personal Communications, Second Edition, Kazimierz Siwiak

Solid Dielectric Horn Antennas, Carlos Salema, Carlos Fernandes, and Rama Kant Jha

Switched Parasitic Antennas for Cellular Communications, David V. Thiel and Stephanie Smith

Understanding Electromagnetic Scattering Using the Moment Method: A Practical Approach, Randy Bancroft

WIPL-D: Electromagnetic Modeling of Composite Metallic and Dielectric Structures, Software and User's Manual, Branko M. Kolundzija, et al.

For further information on these and other Artech House titles, including previously considered out-of-print books now available through our In-Print-Forever® (IPF®) program, contact:

Artech House
685 Canton Street
Norwood, MA 02062
Phone: 781-769-9750
Fax: 781-769-6334
e-mail: artech@artechhouse.com

Artech House
46 Gillingham Street
London SW1V 1AH UK
Phone: +44 (0)20 7596-8750
Fax: +44 (0)20 7630 0166
e-mail: artech-uk@artechhouse.com

Find us on the World Wide Web at:
www.artechhouse.com